ARM-BASED MICROCONTROLLER PROJECTS USING MBED

ARM-BASED MICROCONTROLLER PROJECTS USING MBED

Dogan Ibrahim

Newnes is an imprint of Elsevier
The Boulevard, Langford Lane, Kidlington, Oxford OX5 1GB, United Kingdom
50 Hampshire Street, 5th Floor, Cambridge, MA 02139, United States

Notices
Knowledge and best practice in this field are constantly changing. As new research and experience broaden our understanding, changes in research methods, professional practices, or medical treatment may become necessary.

Practitioners and researchers must always rely on their own experience and knowledge in evaluating and using any information, methods, compounds, or experiments described herein. In using such information or methods they should be mindful of their own safety and the safety of others, including parties for whom they have a professional responsibility.

To the fullest extent of the law, neither the Publisher nor the authors, contributors, or editors, assume any liability for any injury and/or damage to persons or property as a matter of products liability, negligence or otherwise, or from any use or operation of any methods, products, instructions, or ideas contained in the material herein.

Library of Congress Cataloging-in-Publication Data
A catalog record for this book is available from the Library of Congress

British Library Cataloguing-in-Publication Data
A catalogue record for this book is available from the British Library

ISBN: 978-0-08-102969-5

For information on all Newnes publications
visit our website at https://www.elsevier.com/books-and-journals

Working together
to grow libraries in
developing countries

www.elsevier.com • www.bookaid.org

Publisher: Mara Conner
Acquisition Editor: Tim Pitts
Editorial Project Manager: Leticia M. Lima
Production Project Manager: Nirmala Arumugam
Cover designer: Mark Rogers

Typeset by SPi Global, India

Contents

12. SPI Bus Projects

13. UART Projects

14. Advanced Projects

15. Mbed RTOS Projects

16. Internet of Things (IoT)

17. STM32 Nucleo Expansion Boards

About the Author

Dogan Ibrahim has First Class Honors BSc degree in electronic engineering, an MSc degree in automatic control engineering, and a PhD degree in digital signal processing, all obtained from the universities in the United Kingdom. He has worked in many industrial organizations many years before he returned to academic life. He is the author of over 70 technical books and over 200 technical articles and conference proceedings on microcontrollers, microprocessors, digital control systems, and related fields. He is currently a retired chartered electrical engineer and a fellow of the Institution of the Engineering Technology (IET) in United Kingdom. His hobbies include reading, playing tennis, walking, and watching the television.

Preface

Traditionally, a computer was built using a microprocessor chip and many external support chips. A microprocessor includes the central processing unit (CPU), arithmetic and logic unit (ALU), and the timing and control circuitry, and as such it is not useful on its own. A microprocessor must be supported by many external chips such as memory, input-output, timers, interrupt circuits, etc. before it becomes a useful computer. The disadvantage of this type of design was that the chip count was large, resulting in complex design and wiring and high-power consumption.

A microcontroller on the other hand is basically a single chip computer including a CPU, memory, input-output circuitry, timers, interrupt circuitry, clock circuitry, and several other circuits and modules, all housed in a single silicon chip. Early microcontrollers were very limited in their capacities and speed and they consumed considerably more power. Most of the early microcontrollers were 8-bit processors with clock speeds in the range of several MHz, and having only hundreds of bytes of program and data memories. These microcontrollers were traditionally programmed using the assembly languages of the target processors. The 8-bit microcontrollers are still in common use, especially in small projects where large amounts of memory or high speed are not the main requirements. With the advancement of the chip technology we now have 32-bit and 64-bit microcontrollers with speeds in the range of several GHz, and having over several GB of memories. Microcontrollers are nowadays programmed using a high-level language such as C, C#, BASIC, PASCAL, JAVA, etc.

The ARM processors are a family of CPUs based on the reduced instruction set computer (RISC) architecture, developed by advanced RISC machines (ARM). These processors are 32-bit and 64-bit optimized multicore chips, designed for higher speed and lower power consumption than the traditional complex instruction set computers (CISC). Currently, ARM processors are used in electronic devices, such as smartphones, tablets, games consoles, multimedia players, and in many other consumer devices. The ARM processors offer advantages of small size, fast throughput, low cost, and low power consumption. Consequently, most of the smartphones nowadays use a type of ARM processor.

Many electronic companies offer ARM-based processors with different overall architectures and speeds. This book is about using the Nucleo family of microcontrollers in projects. These are 32-bit ARM-based processors offering very high throughputs, large data and program memories, large number of input and output pins, peripheral support modules such as universal asynchronous receiver/transmitter (UART), I^2C bus, serial peripheral interface (SPI) bus, pulse width modulation (PWM) support, timers, external and internal vectored interrupt support, analog-to-digital converters, digital-to-analog converters, and many other features. The book has been organized as a project book where the main theme has been to teach the use of the Nucleo microcontrollers by doing projects.

All the projects in the book have been programmed using the Mbed integrated development environment. Broadly, the book is divided into three parts: basic projects, intermediate projects, and advanced projects. The projects have been organized in the sequence of increasing complexity and therefore it is recommended that the readers complete the early projects before going into the more complex ones. All the projects in the book have been tested and are working and the complete circuit diagrams and full program listings of all the projects are given with detailed descriptions of every line of the programs.

The projects in the book are organized with the following subheadings:

- Title of the project
- Aim of the project
- Block diagram of the project
- Circuit diagram of the project
- Construction of the project (where necessary)
- Program description language (PDL) of the project
- Program listing of the project
- Suggestions for additional work (where necessary)

The popular Nucleo-F411RE-type microcontroller development board is used in almost all the projects in the book (a few of the projects use the Nucleo-L476RG development board). This is a low-cost, high-speed ARM-based microcontroller manufactured by STMicroelectronics. The author has attempted to cover most features of the Nucleo-F411RE microcontroller and Mbed in various projects. In particular, the following features are covered:

- general purpose input-output ports (GPIO)
- analog-to-digital converters (ADC)
- digital-to-analog converters (DAC)
- timers
- external and internal (timer) interrupts
- pulse width modulation (PWM)
- I^2C bus
- SPI bus
- UART
- PWM
- real-time operating system (RTOS)
- Bluetooth
- Wi-Fi

Communication is nowadays one of the most important topics in the computer technology. The book includes several projects on using Wi-Fi and Bluetooth to communicate with the development board either from a smartphone or from a PC.

The RTOS and multitasking are necessary programming tools in complex large real-time projects. Mbed multitasking features are explained in detail in the book and several projects are given on using these features.

Internet of Things (IoT) has become an important monitoring and control application of the microcontrollers. It is estimated that there were over 9 billion IoT devices in use in 2017 and this number is increasing every year. The book explains the basic principles and various architectures of the IoT. Additionally, a complete home-based IoT project is given in the book that uses three development boards with each having a Bluetooth interface and communicating with a PC.

The author hopes that you find the book enjoyable to read and it becomes a source of information for your next projects on microcontrollers.

Dogan Ibrahim

Acknowledgments

The following figures and pictures in this book are taken from the STMicroelectronics source: **UM1724 User Manual, STM32 Nucleo-64 Boards (DocID025833 Rev 12)**.

Figs. 4.1–4.4, 5.1–5.8, 6.1–6.3, 6.5, 6.6.

The pictures of Nucleo Expansion Boards in Chapters 14 and 17 of the book are taken from the following STMicroelectronics internet source: http://www.st.com/en/evaluation-tools/stm32-nucleo-expansion-boards.html?querycriteria=productId=SC1971.

Pictures of Expansion Board pin outs in Chapter 17, Figs. 7.5–7.8, 14.6, and Figs. A.2–A.5 are taken from the Mbed internet sources such as: https://os.mbed.com/components/X-NUCLEO-IDW01M1/.

All of the foregoing figures/pictures are used with written permission of:

©**STMicroelectronics. Used with permission**.

Fig. 3.2 is **Reproduced with permission from Arm Limited. Copyright© Arm Limited (or its affiliates)**.

Fig. 3.3 is taken from the following web page with the permission of Jonathan Valvano: http://users.ece.utexas.edu/~valvano/Volume1/E-Book/C2_FundamentalConcepts.htm.

Figs. 10.16 and 10.18 are taken from the mikroElektronika web site: www.mikroe.com.

Figs. 11.2 and 11.3 are taken from the SparkFun Electronics web site: www.sparkfun.com, copied under Creative Commons license (https://creativecommons.org/licenses/by/2.0/).

Figs. 14.3 and 14.4 are taken from the Elegoo web site: www.elegoo.com.

Figs. 8.60 and 8.61 are taken from Analog Devices Data sheet: Low Voltage Temperature Sensors, TMP35/TMP36/TMP37, Rev: H, Doc: D00337-0-5/15(H).

The author would like to thank to STMicroelectronics, ARM Limited (or its affiliates), mikroElektronika, SparkFun Electronics, Elegoo, and Analog Devices Inc for giving permission to use the foregoing pictures/figures/tables in this book. The author is also grateful to Michael Markowitz of STMicroelectronics for providing sample Nucleo processor boards and Nucleo Expansion boards for use in the projects in this book.

Introduction

1.1 OVERVIEW

Microcontrollers are single chip computers that include a minimum of a microprocessor, memory, and input-output module. Depending on the complexity, some microcontrollers include additional components such as counters, timers, interrupt control circuits, serial communication modules, analog-to-digital converters, digital signal processing modules, and so on. Thus, a microcontroller can be anything from a tiny single chip embedded controller to a large computer system having keyboard, monitor, hard disk, printer, and so on.

A microprocessor is different from a microcontroller in many different ways. The main difference is that a microprocessor requires several additional external support chips such as memory and input-output circuits before it can be used as a digital controller. A microcontroller on the other hand includes all these support chips on the same chip and that is why it is called a single chip computer. As a result, multiple chip microprocessor-based computer systems consume considerably more power than microcontroller-based systems. The costs of the single chip microcontroller systems are also much lower than the costs of the multiple chip-based microprocessor systems.

Microprocessors and microcontrollers operate by executing user programs. These programs are stored in the program memory of the device and consist of instructions that can be understood and obeyed by the device. The device fetches these instructions from its program memory one by one and then implements the required operations. Under the control of the user program data is received from external input devices (inputs), manipulated as requested, and then sent to external devices (outputs).

Microcontrollers (and microprocessors) have traditionally been programmed using the assembly language of the target device. The assembly language consists of many mnemonics where each mnemonic describes a basic instruction that can be carried out by the device. Although the assembly language is very fast, it has many disadvantages. Firstly, because of the syntax of the assembly language, it is difficult to learn this language. Secondly, processors developed by different manufacturers have different sets of assembly language instructions. Even in most cases the processors manufactured by the same manufacturer may have different assembly language instruction sets. As a result, the programmer may be required to learn

1

a different assembly language every time a new processor is to be used. Thirdly, in general, it is difficult to maintain a program written using the assembly language.

Although the assembly language is still in use in some real-time applications, nowadays most applications are developed using a high-level language, such as BASIC, C, C++, C#, Visual BASIC, PASCAL, JAVA, and so on. Perhaps the greatest advantage of the high-level languages is the ease of learning and fast program development. Large and complex programs can be developed in much shorter times compared to the assembly language. For example, to write a piece of assembly language program code to multiply two floating point numbers can take several hours or even more time and mistakes can easily be made. On the other hand, using a high-level language we just multiply the two numbers. Additionally, it is much easier to maintain a program written using a high-level language. High-level languages also have the benefit that in general the same user program can easily be transported to work on a different processor with little or no modifications. High-level languages are supported by large number of built-in libraries that make it easy to develop very complex programs in relatively short times. Finally, another advantage of using the high-level languages is that the developed program can easily be tested and this feature shortens the development time considerably.

In this book we shall be using the C language which is perhaps currently the most popular language used in microcontroller-based applications. As we shall see in the later chapters, the Mbed integrated development environment will be used to develop our projects. Mbed is an online integrated development environment that is used to develop ARM (Advanced RISC Machines) processor-based applications. Using Mbed we can write a program in C, then compile the program, and upload the executable code to the target ARM processor. The advantage of using Mbed is that it is easy to learn and use and is supported by very large number of library functions. The projects in this book are all based on the ARM processor. There are many ARM development boards available in the market. In this book we shall be using a model of the STM32-Nucleo development boards manufactured by STMicroelectronics. As will be discussed in detail in the later chapters, STM32-Nucleo boards are complete microcontroller development boards incorporating fast 32-bit ARM processors together with all the support circuitry to help develop complex projects. The STM32-Nucleo development boards are Mbed compatible which makes them ideal boards for developing complex applications in relatively short times.

1.2 EXAMPLE MICROCONTROLLER-BASED CONTROL SYSTEM

In this section we shall see how a microcontroller can be used in a simple control system application. Fig. 1.1 shows a liquid control system where the aim is to control the level of the liquid in the reservoir at a specified point. Water is pumped from the reservoir to the tank using a pump and pipes. In Fig. 1.1 the level of the liquid is controlled manually without using a microcontroller. Here, the person in charge observes the liquid level inside the tank and turns the pump off when the liquid level reaches the required prespecified level.

The system shown in Fig. 1.1 is manual and requires constant attention of a person. A simple microcontroller version of this system is shown in Fig. 1.2. Here, the liquid level is read by

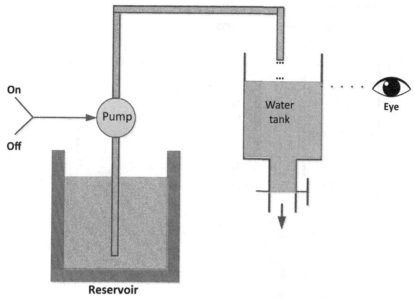

FIG. 1.1   Manual liquid level control system.

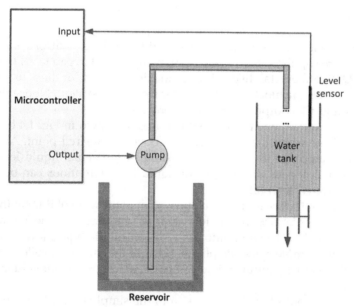

FIG. 1.2   Microcontroller-based liquid level control system.

the microcontroller via a liquid level sensor device. The program running inside the microcontroller compares the actual liquid level with the desired level and then actuates the pump
automatically in order to keep the liquid at the desired level. If the liquid inside the tank is
low, the microcontroller operates the pump to draw more liquid from the reservoir.

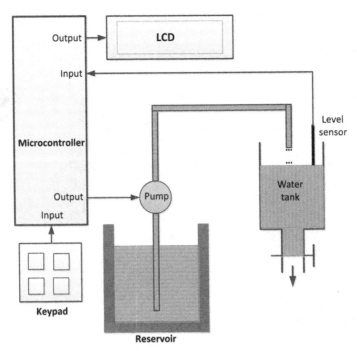

FIG. 1.3   Adding a keypad and an LCD.

The system shown in Fig. 1.2 is a very simplified liquid level control system with no user interaction. In a more sophisticated system we may include a keypad to set the desired liquid level and an LCD (liquid crystal display) to see the desired and/or the actual liquid levels in the tank. Fig. 1.3 shows the block diagram of our upgraded system. Notice that here we are using two inputs and two outputs from our microcontroller.

We can make our system even more sophisticated as shown in Fig. 1.4 by adding an audible alarm to indicate when the water level is above the desired point. Also, a PC can be interfaced to the microcontroller so that, for example, the actual liquid levels can be sent to the PC at regular intervals and graphs of liquid level variations can be plotted on the PC between the required intervals.

In Fig. 1.5, wireless interface is added to our system in the form of Bluetooth or Wi-Fi. With the help of the wireless interface we can, for example, send and save the liquid level readings on a Cloud. Additionally, we can monitor and/or control the liquid level remotely through the Cloud using, for example, a mobile phone. Because the microcontrollers are programmable and in general offer many input and output ports, we can make our system as simple or as complex as we like.

The power of microcontrollers is obvious from the simple example given in this section. Microcontrollers are classified by the number of bits they process at a time. Although some of the early microcontrollers were only 4 bits, the 8-bit devices are still the most popular and commonly used devices. Examples of some 8-bit microcontrollers are PIC16, PIC18, Arduino, 8051, and so on. The 16-bit and 32-bit microcontrollers are faster, have more memories, and are more powerful, but at the same time they are more expensive and their use may not be

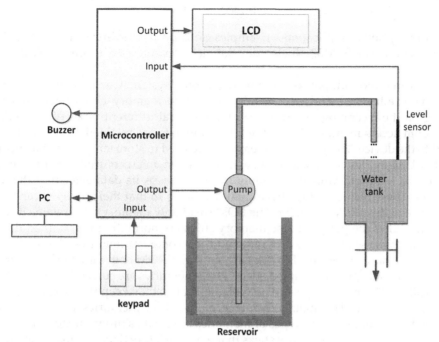

FIG. 1.4 Adding an audible alarm and a PC.

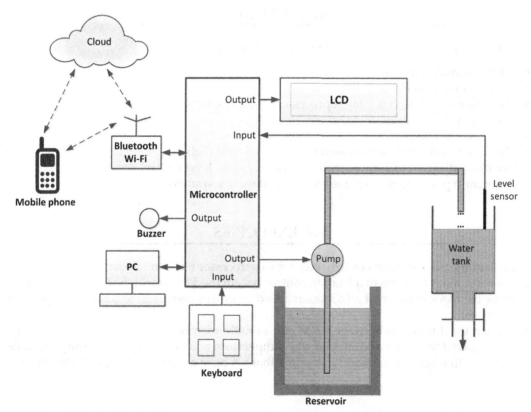

FIG. 1.5 Adding wireless interface.

justified in many small applications. Examples of 32-bit microcontrollers are PIC32, ARM family, and so on. The STM32-Nucleo development boards used in this book incorporate 32-bit processors.

Memory is an important part of any microcontroller system. Currently the memory technology is very advanced and as a result the cost of memory chips have come down. Depending on the technology used we can have several different types of memories. The RAM (random access memory) is volatile where the data are retained as long as the power is applied to the device. These types of memories are used to store temporary data in our programs. RAM memories are used as data memories in microcontroller applications. The EPROM (erasable programmable read-only memory) keeps its data even after the removal of the power. These memory chips have small windows so that their contents can be erased with exposure to the ultraviolet light. The ROM (read only memory) also retains its contents after the removal of the power. These memory chips are normally programmed during the manufacturing process and once programmed their contents cannot be changed. The programmable read-only memories (PROMs) are similar to ROMs but they can be programmed once by using suitable programming devices. Flash memories are currently used in almost all microcontrollers. These memories are not volatile, that is, they keep their data even after the removal of the power. The main advantage of the flash memories is that they can be programmed many times. These types of memories are used as program memories in microcontroller systems. The USB memory sticks that we use to store data also use flash memories.

1.3 SUMMARY

In this chapter we have learned about the following:

- What is a microprocessor?
- What is a microcontroller?
- The differences between a microprocessor and a microcontroller.
- The assembly language.
- High-level languages.
- The differences between the assembly language and high-level languages.
- The block diagram of a microcontroller-based liquid level control system.
- Different types of memories used in microcontroller systems.

1.4 EXERCISES

1. Explain the main differences between a microprocessor and a microcontroller.
2. Draw the block diagram of a heating control system used to control the heating in a room.
3. Draw the block diagram of a DC motor speed control system used to control the speed of the motor.
4. Explain the differences between RAM and EPROM memories.
5. It is required to control and monitor the temperature of an oven remotely using a mobile phone with Bluetooth capability. Assume that we have available a temperature sensor

chip, a heater, a fan, and a Bluetooth interface module. Draw the block diagram to show how the system may be configured.

6. Repeat Exercise 5 by assuming that we wish to control and monitor the temperature with our mobile phone and Wi-Fi by sending and receiving data through the Cloud. Assume that we already have a local Wi-Fi router and a Wi-Fi interface module for our microcontroller.

7. In Exercise 6, explain the advantage of monitoring/controlling our oven through the Cloud.

8. Explain how the system block diagram in Exercise 5 should change if we wish to monitor the oven temperature via SMS messages received on our mobile phone.

Microcontroller-Based Project Development Cycle

2.1 OVERVIEW

Microcontroller-based project development cycle has two stages: program development cycle (PDC) and hardware development cycle (HDC). Both development cycles are based around simple loops. In this section we briefly look at both the development cycles.

2.1.1 The Program Development Cycle (PDC)

Fig. 2.1 shows the program development cycle (PDC) which basically consists of developing the application program. This is usually carried out using a development board (or a development kit) incorporating the target microcontroller and additional components and interface devices (e.g., transistors, LEDs, motors, relays, sensors, etc.) that may be required for the project. At the beginning of the PDC we write the source code using the in-built text editor of the chosen integrated development environment (IDE). We then compile the source code and remove any syntax errors or any other compile time errors. Notice that at this stage we are not sure whether our source code is logically correct or not. At this stage we can either simulate or debug our source code using the tools offered by the IDE. Using the simulator has the advantage that we can test our source code on our PC without having to upload the executable code to the program memory of the target microcontroller. With the help of the simulator we can single step through our source code, examine the variables, insert breakpoints, modify the variables, and so on. Simulators are available almost on all IDEs and they are very easy to use. Simulation helps the programmer to verify that the source code may be logically correct and the application is probably working. Of course simulation is not same as the real testing using the target hardware. There still could be logical errors and the application may not behave as desired and such errors will not be shown during the simulation stage. If any errors are detected at the simulation stage we should go back and correct the source code to remove these errors. After a successful simulation the next stage in the PDC cycle is to upload the executable code to the program

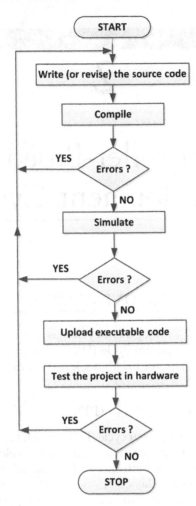

FIG. 2.1   Program development cycle.

memory of the target microcontroller (on the development board). The program is then tested on the real hardware to ensure that the project is working as desired. Enough time should be spent at this stage in order to find out if there are any application errors. If any errors are detected then we should go back to the beginning of the cycle and correct our source code to remove these errors. The above loop should be repeated until all the errors are removed from the project and we are completely satisfied with the correctness of the project.

2.1.2 The Hardware Development Cycle (HDC)

A typical hardware development cycle (HDC) is shown in Fig. 2.2. After testing our project with the developed source code and using the microcontroller development board together with the required support components and interfaces devices, the next stage of the project is

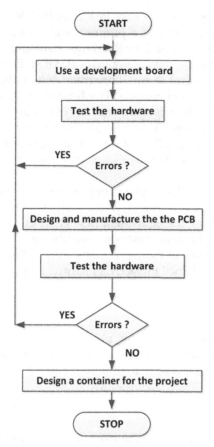

FIG. 2.2 Hardware development cycle.

to build the hardware on a printed circuit board (PCB). The complete circuit diagram is designed using an integrated schematic and PCB design software. The size of the final PCB is determined and the target microcontroller chip and all the other required components are placed on the PCB board using the PCB design software. The auto-router option is then used to interconnect all the components as is shown in the schematic. The next stage is the manufacturing of sample PCB boards which is normally done by sending the necessary PCB files to a qualified PCB manufacturer. The executable code is uploaded to a target microcontroller chip and all the components are then mounted and soldered on the PCB board (this step can also be done by the PCB manufacturers). The prepared hardware is then tested together with any external devices. If the project is successful the final stage is usually the design and manufacturing of a suitable container for the project.

2.2 PROGRAM DEVELOPMENT TOOLS

Simple programs consisting of no more than 10 lines of code can easily be developed by writing the code without any prior preparation. The development of large and complex programs is easier if an algorithm is first derived and the program is broken down into smaller modules. The actual coding becomes an easy task after the algorithm is available. An algorithm describes the operational steps of a program and it can be in the form of graphical or text based, such as flow charts, data flow diagrams, structure charts, program description languages (PDLs), and unified modeling languages (UMLs). Flow charts can be very useful tools to describe the flow of control in small programs where there are no more than a few pages of diagrams. The problem with graphical tools is that it can be time consuming to draw or to modify them, especially if there is more than one diagram extending over several pages. PDL is not a programming language. It is a collection of text-based keywords and actions that help the programmer to describe the flow of control and data in a program in a stepwise and logical manner. The main advantage of the PDL is that it is very easy to modify a given PDL since it only consists of text.

In this book we shall be using PDLs wherever possible, and flow charts will also be given where it is felt to be useful. In the next section we shall be looking at the basic constructs of PDL and at the same time show the equivalent flow chart of each PDL construct.

Note: There are many free of charge programs available on the Internet that can be used to help draw flow chart easily. Some of these programs are as follows: Microsoft Visio, Dia, yEd Graph Editor, ThinkComposer, Pencil Project, LibreOffice, Diagram Designer, LucidChart, etc.

2.2.1 BEGIN-END

Every PDL description must start with a BEGIN and terminate with an END. The keywords should be in bold and the statements inside these keywords should be indented to make the reading easier. An example is shown in Fig. 2.3.

2.2.2 Sequencing

In normal program flow, statements are executed in sequence one after the other. The operations to be performed in each step are written in plain text. An example of sequencing is shown in Fig. 2.4 together with its flow chart equivalent.

2.2.3 IF-THEN-ELSE-ENDIF

The IF-THEN-ELSE-ENDIF statements are used to create conditional statements and thus to change the flow of control in a program. Every IF statement must be terminated with an ENDIF statement. The ELSE statement is optional and if used it must be terminated with an ENDIF statement. It is also permissible to use ELSE IF statements in programs where multiple decisions have to be made. Figs. 2.5–2.7 show various examples of using the IF-THEN-ELSE-ENDIF statements.

FIG. 2.3   BEGIN-END statement and its equivalent flow chart.

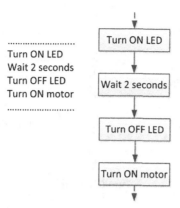

FIG. 2.4   Sequencing and its equivalent flow chart.

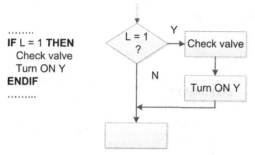

FIG. 2.5   Using IF-THEN-ENDIF.

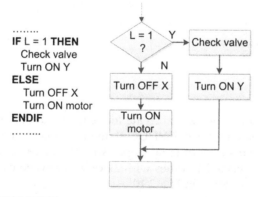

FIG. 2.6   Using IF-THEN-ELSE-ENDIF.

2.2.4 DO-FOREVER-ENDDO

The DO-FOREVER-ENDDO statement is used to repeat a loop forever. This kind of loop is commonly used in microcontroller applications where an operation or a number of operations are executed continuously. Fig. 2.8 shows an example of using the DO-FOREVER-ENDDO statement.

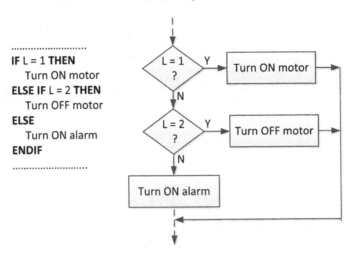

```
........................
IF L = 1 THEN
    Turn ON motor
ELSE IF L = 2 THEN
    Turn OFF motor
ELSE
    Turn ON alarm
ENDIF
........................
```

FIG. 2.7   Using IF-THEN-ELSE IF-ENDIF.

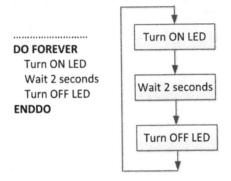

```
.......................
DO FOREVER
    Turn ON LED
    Wait 2 seconds
    Turn OFF LED
ENDDO
```

FIG. 2.8   Using DO-FOREVER-ENDDO statement.

2.2.5 DO-ENDDO

The DO-ENDDO statement is used to create loops (or iterations) in programs. Every Do statement must be terminated with an ENDDO statement. It is permissible to use conditions after the DO statement to create conditional loops. An example DO-ENDDO loop is shown in Fig. 2.9 where the LED is flashed 10 times with 2s delay between each output. Another example DO-ENDDO loop is shown in Fig. 2.10.

2.2.6 REPEAT-UNTIL

The REPEAT-UNTIL statement is similar to DO-ENDDO statement but here the condition to terminate the loop is checked at the end and therefore the loop is executed at least once. Fig. 2.11 shows an example of REPEAT-UNTIL loop.

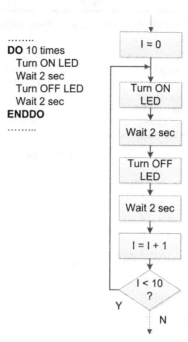

```
........
DO 10 times
   Turn ON LED
   Wait 2 sec
   Turn OFF LED
   Wait 2 sec
ENDDO
........
```

FIG. 2.9 Using DO-ENDDO statement.

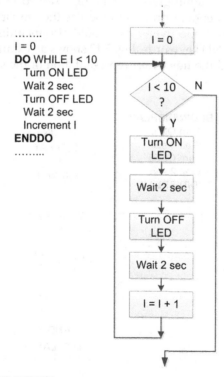

```
........
I = 0
DO WHILE I < 10
   Turn ON LED
   Wait 2 sec
   Turn OFF LED
   Wait 2 sec
   Increment I
ENDDO
........
```

FIG. 2.10 Another example DO-ENDDO statement.

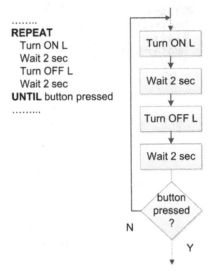

....
REPEAT
   Turn ON L
   Wait 2 sec
   Turn OFF L
   Wait 2 sec
UNTIL button pressed
....

FIG. 2.11   Using REPEAT-UNTIL statement.

2.2.7 Subprograms

There are many ways that subprograms can be represented in PDLs and in flow charts. Since subprograms are independent program modules they must start and finish with the BEGIN and END statements, respectively. We should also include the name of the subprogram after the BEGIN and END keywords. Fig. 2.12 shows an example of subprogram called DISPLAY. Both the PDL and the flow chart representation of the subprogram are shown in this figure.

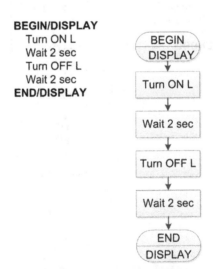

BEGIN/DISPLAY
   Turn ON L
   Wait 2 sec
   Turn OFF L
   Wait 2 sec
END/DISPLAY

FIG. 2.12   Subprogram DISPLAY.

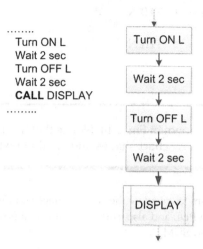

Turn ON L
Wait 2 sec
Turn OFF L
Wait 2 sec
CALL DISPLAY

FIG. 2.13   Calling subprogram DISPLAY.

2.2.8 Calling a Subprogram

A subprogram can be called from the main program or from another subprogram. In PDLs, a subprogram is called by using the keyword CALL, followed by the name of the subprogram. In flow charts it is common practice to insert vertical lines at the two sides of the box where the subprogram is called. Fig. 2.13 shows an example of where subprogram DISPLAY is called from a program.

2.3 EXAMPLES

Some simple examples are given in this section to show how the PDLs and their equivalent flow charts can be used in program development.

EXAMPLE 2.1

It is required to write a program to convert hexadecimal numbers "A" to "F" into decimals. Show the algorithm using a PDL and also draw the equivalent flow chart. Assume that the hexadecimal number to be converted is called HEX_NUM, and the output decimal number is called DEC_NUM.

Solution 2.1

The required PDL is:

```
BEGIN
    IF HEX_NUM = "A" THEN
        DEC_NUM = 10
    ELSE IF HEX_NUM = "B" THEN
        DEC_NUM = 11
    ELSE IF HEX_NUM = "C" THEN
        DEC_NUM = 12
    ELSE IF HEX_NUM = "D" THEN
        DEC_NUM = 13
```

```
        ELSE IF HEX_NUM = "E" THEN
                DEC_NUM = 14
        ELSE IF HEX_NUM = "F" THEN
                DEC_NUM = 15
        ENDIF
END
```

The required flow chart is shown in Fig. 2.14. Notice that it is much easier to write the PDL statements than drawing the flow chart shapes and writing text inside them.

EXAMPLE 2.2

It is required to write a program to calculate the sum of integer numbers between 1 and 100. Show the required algorithm using a PDL and also draw the equivalent flow chart. Assume that the sum will be stored in variable called SUM.

Solution 2.2

The required PDL is:

```
BEGIN
        SUM = 0
        I = 1
        DO 100 TIMES
                SUM = SUM + I
                Increment I
        ENDDO
END
```

The required flow chart is shown in Fig. 2.15.

EXAMPLE 2.3

An LED is connected to a microcontroller output port. Additionally a button is connected to one of the input ports. It is required to turn ON the LED when the button is pressed, and otherwise to turn OFF the LED. Show the required algorithm using a PDL and also draw the equivalent flow chart.

Solution 2.3

The required PDL is:

```
BEGIN
        DO FOREVER
                IF Button is pressed THEN
                        Turn ON LED
                ELSE
                        Turn OFF LED
                ENDIF
        ENDDO
END
```

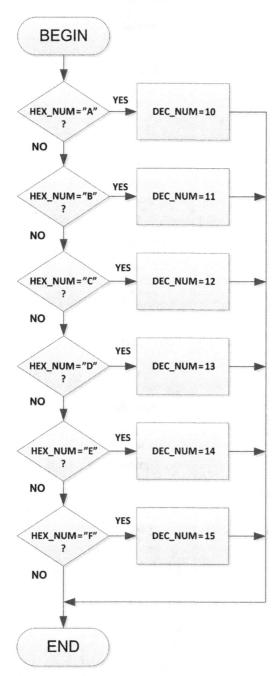

FIG. 2.14   Flow chart solution.

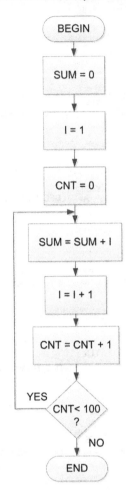

FIG. 2.15   Flow chart solution.

The required flow chart is shown in Fig. 2.16.

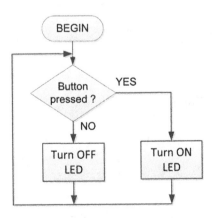

FIG. 2.16   Flow chart solution.

EXAMPLE 2.4

An analog pressure sensor is connected to one of the analog-to-digital (ADC) input ports of a microcontroller. Additionally, an LCD display is connected to the output ports. It is required to read the pressure every second and then display on the LCD. Show the required algorithm using a PDL and also draw the equivalent flow chart. Assume that the display routine is a subprogram called DISPLAY.

Solution 2.4

The required PDL is:

BEGIN
    DO FOREVER
        Read Temperature from ADC port
        CALL DISPLAY
        Wait 1 second
    ENDDO
END
BEGIN/DISPLAY
    Display temperature on LCD
END/DISPLAY

The required flow chart is shown in Fig. 2.17.

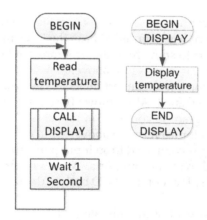

FIG. 2.17   Flow chart solution.

2.4 SUMMARY

In this chapter we have learned about the following:

- Program development cycle
- Hardware development cycle

- Program development tools
- PDLs
- Flow charts

2.5 EXERCISES

1. What are the advantages and disadvantages of flow charts?
2. It is required to write a program to calculate the sum of even integer numbers from 1 to 10. Draw the flow chart to show the algorithm for this program.
3. Write the PDL statements for exercise (2) above.
4. It is required to write a function to calculate the area of a cylinder given its radius and height. Draw flow charts to show how this function can be implemented. Also, draw a flow chart to show how the function can be called from a main program to calculate the area of a cylinder whose height is 2.5 cm and its radius is 3 cm.
5. Draw the equivalent flow chart for the following PDL statements:

> **BEGIN**
>> **DO** 10 times
>>> Turn ON relay
>>> Wait 1 second
>>> Turn OFF relay
>>> Wait 1 second
>> **ENDO**
> **END**

6. A push-button switch is connected to a microcontroller input ports. Also, an LED is connected to an output port. It is required to flash the LED three times when the button is pressed. Draw a flow chart to show how the program can be designed.
7. Write the PDL statements for exercise (6) above.
8. It is required to write a function to convert meters to inches. Draw a flow chart to show how this function can be designed. Also, draw a flow chart to show how the function can be called from a main program to convert 5 m to inches.
9. Write the PDL statements for exercise (8) above.
10. A temperature sensor chip is connected to an input port of a microcontroller. At the same time an LCD is connected to the output ports. Write a program to read the temperature every minute and then display it on the LCD both in degrees Fahrenheit and in degrees Centigrade.
11. Draw the flow chart equivalent of the following PDL:

> **BEGIN**
>> Configure the output port
>> **IF** switch 1 is pressed **THEN**
>>> Start the motor
>> **ELSE IF** switch 2 is pressed **THEN**
>>> Start the pump
>> **ELSE**
>>> Turn ON the LED
>> **ENDIF**
> **END**

12. Draw the flow chart equivalent of the following PDL:

      BEGIN
             Wait for switch to be pressed
             Turn ON the buzzer
             Wait 5 seconds
             Turn OFF buzzer
             Wait until switch is released
      END

13. Draw the flow chart equivalent of the following PDL:

      BEGIN
          DO FOREVER
               IF motor is ON **THEN**
                    Turn ON pump
                    Wait for 5 seconds
               ELSE
                    Turn ON alarm
                    Wait for 10 seconds
                    Turn OFF alarm
               ENDIF
          ENDDO
      END

3

The ARM Microcontrollers

3.1 OVERVIEW

In this chapter we shall be looking briefly at the history and architecture of the ARM processors and ARM processor-based development boards.

3.2 BRIEF HISTORY OF ARM PROCESSORS

The history of ARM processor is interesting. In 1981, the British Broadcasting Corporation (BBC) invited the computer firms to supply a computer to be used in educational projects. The Acorn Computers Ltd. was the winner and they developed a home computer based on the 8-bit microcontroller 6502, manufactured by the company MOS Technology. Although the 6502 was not a powerful microprocessor with today's standards, it was fast enough in those days and became very successful in the early 1980s.

The founders of the Acorn Computers Ltd. were Christopher Curry and Herman Hauser. Christopher used to work closely with Clive Sinclair of Sinclair Radionics Ltd. for over a decade. Sinclair had some financial troubles and later formed a new company called Sinclair Research Ltd. where Christopher was one of the main people in this new company. After a disagreement with Sinclair, Christopher left the company and formed the Acorn Computers Ltd with the Austrian physicist Herman Hauser.

During the early and late 1980s IBM produced their first PC based on the 16-bit microprocessor 8088. During these years IBM became very popular in the PC market with the early MSDOS operating system and many small competitors faded. The Acorn 6502 was only 8 bits and it was not powerful enough for graphics-based applications. During this time Acorn decided that they needed a new architecture with faster processor and they considered designing their own processor.

In 1990s Acorn came to the conclusion that their future lied not only in selling computers, but also in developing new computer architectures. Therefore, in 1990 Acorn Computers Ltd. established a new company called Advanced Reduced Instruction Set Computer (RISC) Machines Ltd. (ARM for short) as a joint venture with Apple Computer and VLSI Technology.

In this venture, Apple invested by providing the cash, VLSI provided the necessary technology tools, and Acorn provided the experienced design engineers. By the year ARM became a £26.6 million company with net income of £2.9 million. In 1998 ARM, currently called ARM Holdings, was listed on the London Stock Exchange and NASDAQ. In later years ARM designed new processor architectures and sold the intellectual property rights as licenses to the firms who wished to incorporate ARM designs in their own products. In 2016 SoftBank bought ARM for $31 billion.

ARM has been designing 32-bit processors for over 20 years and in the last few years they have also started to offer 64-bit designs. In actual fact ARM is a company specialized in designing the processor architecture and they do not manufacture or sell processor chips. ARM makes money by licensing their designs to chip manufacturers. The manufacturers use the core ARM processors (e.g., the core CPU) and integrate with their own peripherals to end up in a complete microcontroller chip. ARM is then given royalty fees for each chip manufactured by the third-party companies. Companies that use ARM core processors include Apple, Atmel, Broadcom, Cypress Semiconductors, Freescale Semiconductors, Analog Devices, Nvidia, NXP, Samsung Electronics, Texas Instruments, Qualcomm, Renesas, and many others.

The ARM concept became so popular that in mobile applications in 2005, about 98% of all mobile phones sold used at least one ARM processor. In 2011, the 32-bit ARM architecture was the most widely used and most popular architecture used in mobile devices. Over 10 billion units were produces in 2013. The ARM architecture is known to offer the best MIPS to Watts ratio as well as MIPS to $ ratio in the industry, and the smallest CPU size.

Many ARM cores have been designed over the years, such as the ARMv1, ARMv2, ..., ARMv8 and so on, ranging from 32-bits to 64-bits. The 32-bit ARMv7-A was the most widely used architecture in mobile devices in 2011. This includes architectures such as Cortex-A5, Cortex-A7, Cortex-A8, and so on. Lower performing cores have lower license fees than higher performing cores. The ARMv6-M and ARMv7E-M are used in lower performance microcontroller-based control and monitoring applications and these architectures include Cortex-M0, Cortex-M0+, Cortex-M1, Cortex-M3, Cortex-M4, and Cortex-M7.

3.3 THE ARM PROCESSOR ARCHITECTURE

3.3.1 CICS and RISC

Basically there are two types computer designs as far as the CPU design is considered: *Complex Instruction Set Computer* (CISC) and RISC. Traditionally all CPU designs were based on CISC where the instruction set of these computers had many instructions. Many microprocessors, including the 6502 were based on CISC where some instructions were executed in one cycle and some others required several processor cycles to execute.

Most present day CPUs have limited instructions in their instructions sets and are therefore called *RISC*. ARM processors use RISC-based designs and this is one of the reasons that they are used widely, especially in mobile devices. Because of simpler designs the RISC processors use less number of transistors and therefore reduced power consumption. These processors execute one instruction per cycle and as a result these CPUs are fast, but they may need to execute more instructions for a given task. Normally, each instruction in a RISC-based design occupies one memory location. These instructions are fetched from the memory and are executed in one cycle each as fast as possible. RISC-based CPU designs also allow for several levels of pipelining where the next instructions are fetched from the memory while the current instruction is being executed, thus resulting in higher throughput. Also, because the RISC-based designs are simpler, they consume less power which is extremely important in battery operated mobile applications, such as in mobile phones, tablets, GPS systems, or games consoles. This is perhaps the major reason why the ARM architecture is widely used in mobile devices requiring very fast processing times as well as low power consumptions.

3.3.2 Why ARM?

The choice of a microcontroller for a particular application depends on many factors such as the following:

- cost
- speed
- power consumption
- size
- number of digital and analog input-output ports
- digital input-output port current capacity
- analog port resolution and accuracy
- program and data memory sizes
- interrupt support
- timer support
- USART support
- special bus support (e.g., USB, CAN, SPI, I^{2}C, and so on)
- ease of system development (e.g., programming)
- working voltage

For example, if we need to develop a battery powered portable device such as a mobile phone, a tablet, or a games console then high clock speed as well as long battery life are the main requirements. But if on the other hand, for example, we wish to develop a liquid level control system or a heating control system then high speed or low power consumption are not among the desired requirements. In general, as the clock speed goes up so does the power consumption and as a result a trade-off should be made in choosing a microcontroller for a specific application.

ARM processors are based on an instruction set called *Thumb*. With clever design this instruction set takes 32-bit instructions and compresses them down to 16-bits, thus reduces the hardware size, which also reduces the overall cost and the power consumption. The processor makes use of multistage pipelined architecture that is easier to learn, build, and program.

ARM's core architecture is only a processor and it does not include graphics, input-output ports, USB, serial communication, wireless connectivity, or any other form of peripheral modules. Chip manufacturers build their systems around the ARM core design and this is why different manufacturers offer different types of ARM-based microcontrollers.

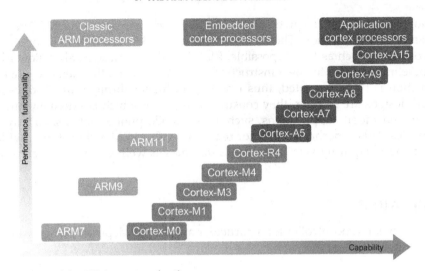

FIG. 3.1   Overview of the ARM processor family.

Over the last 20 years or so ARM had developed many 32-bit processors. Fig. 3.1 shows some of the popular members of the ARM processor family. Around 2003 ARM decided to improve their market share by developing new series of high-performance processors mainly for microcontroller-based applications, such as in embedded control and monitoring applications. As a result, the Cortex family or processors were created. The Cortex family consists of three processor families: Cortex-M, Cortex-R, and Cortex-A. We shall now briefly look at these families.

Cortex-M

Cortex-M series are built around the ARMv6-M architecture (Cortex-M0 and Cortex-M0+) and the ARMv7-M architecture (Cortex-M3 and Cortex-M4). These processors are specifically designed for the microcontroller market, offering quick and deterministic interrupt responses, low power consumption, low cost, fairly high performance, and ease of use. The Cortex-M3 and Cortex-M4 are very similar in architecture and have the same instruction sets (Thumb 2) with the difference that the Cortex-M4 offers digital signal processing (DSP) capabilities and has optional floating point unit (FPU). Cortex-M4 with its DSP and floating point capability is an ideal processor for the IoT and wearable applications. For cost sensitive and lower performance applications the Cortex-M0 or the Cortex-M0+ can be used. The Cortex-M0 processor has small gate count (12K gates) and consumes only 12.5 µW/MHz. The Cortex-M0+ consumes only 9.85 µW/MHz and is based on a subset of the Thumb 2 instruction set and its performance is slightly above that of Cortex-M0 and below that of the Cortex-M3 and Cortex-M4. Cortex-M7 is a high-performance processor that can handle fast DSP and single or double precision floating point operations and is mainly used in applications where higher performance than the Cortex-M4 is required. In this book we will be using a microcontroller development board based on the Cortex-M4 architecture.

Cortex-R

Cortex-R series are real-time higher performance processors than the Cortex-M and some members are designed to operate at high clock rates in excess of 1 GHz. These processors are commonly used in hard-disk controllers, network devices, automotive applications, and specialized in high-speed microcontroller applications. Cortex-R4 and Cortex-R5 are the early members and can be used at clock speeds of up to 600 MHz. Cortex-R7 is a newer member that incorporates 11-stage pipeline for high performance and it can operate in excess of 1 GHz. Although the Cortex-R processors are high performance, their architecture is complex and these processors consume high power, making them unsuitable for use in mobile battery powered devices.

Cortex-A

Cortex-A are the highest performance ARM processors designed for use with real-time operating systems in mobile applications such as in mobile phones, tablets, GPS devices, games consoles, and so on. These processors support advanced features for operating systems such as Android, ioS, Linux, Windows, etc. In addition, advanced memory management is supported with virtual memory. Early members of the family included processors such as Cortex-A5 to Cortex-A17, based on the ARMv7-A architecture. Latest members of the family are the Cortex-A50 and Cortex-A72 series designed for low-power and very high-performance mobile applications. These processors are built using the ARMv8-A architecture which offers 64-bit energy-efficient operation with the capabilities of more than 4 GB of physical memory.

3.3.3 Cortex-M Processor Comparison

A comparison of the various Cortex-M series processors is given in Table 3.1. As can be seen from this table, Cortex-M0 and Cortex-M0+ are used at low speed and low power consumption applications. Cortex-M1 is optimized for use in programmable gate array applications. Cortex-M3 and Cortex-M4 are medium-power processors used in microcontroller applications with the Cortex-M4 supporting DSP and floating point arithmetic operations. Cortex-M7 is a high-performance member of the family which is used in applications requiring higher performance than the Cortex-M4.

TABLE 3.1   Cortex-M Processor Comparison

Processor	Description
Cortex-M7	High-performance processor, used in applications where Cortex-M4 is not fast enough, supports DSP and single and double precision arithmetic
Cortex-M4	Similar architecture as the Cortex-M3 but includes DSP and floating point arithmetic, used in high-end microcontroller-type applications
Cortex-M3	Very popular, low power consumption, medium performance, debug features, used in microcontroller-type applications
Cortex-M1	Designed mainly for programmable gate array applications
Cortex-M0+	Lower power consumption and higher performance than the Cortex-M0
Cortex-M0	Low power consumption, low to medium performance, smallest ARM processor

3.3.4 Processor Performance Measurement

Processor performance is usually measured using benchmark programs. There are many benchmark programs available and one should exercise care when comparing the performance of various processors as the performance depends on many external factors such as the efficiency of the compiler used and the type of operation performed for the measurement.

Many attempts were made in the past to measure the performance of a processor and quote it as a single number. For example, MOPS, MFLOPS, Dhrystone, DMIPS, BogoMIPS, and so on. Nowadays, CoreMark is one of the most commonly used benchmark programs used to indicate the processor performance. CoreMark is developed by Embedded Microprocessor Benchmark Consortium (EEMBC, www.eembc.org/coremark) and is one of the most reliable performance measurement tools available.

Table 3.2 shows the CoreMark results for some of the commonly used microcontrollers. As can be seen from this table, Cortex-M7 achieves 5.01 CoreMark/MHz, while the PIC18 microcontroller achieves only 0.04 CoreMark/MHz.

3.3.5 Cortex-M Compatibility

Processors in the Cortex family are upward compatible with each other. Cortex-M0 and Cortex-M0+ processors are based on the ARMv6-M architecture, using the Thumb instruction set. On the other hand, Cortex-M3, Cortex-M4, and Cortex-M7 are based on the ARMv7-M architecture, using the Thumb 2 instruction set which is a superset of the Thumb instruction set. Although the architectures are different, software developed on the Cortex-M0 and Cortex-M0+ processors can run on the Cortex-M3, Cortex-M4, and Cortex-M7 processors without any modifications provided the required memory and input-output ports are available.

TABLE 3.2   CoreMark/MHz for Some Commonly Used Microcontrollers

Processor	CoreMark/MHz
Cortex-M7	5.01
Cortex-A9	4.15
Cortex-M4	3.40
Cortex-M3	3.32
Cortex-M0+	2.49
Cortex-M0	2.33
dsPIC33	1.89
MSP430	1.11
PIC24	1.88
PIC18	0.04

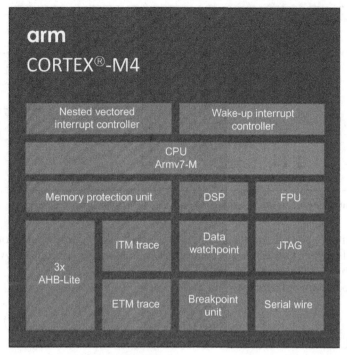

FIG. 3.2   Simplified block diagram of the Cortex core.

3.3.6 The Cortex Core Architecture

Fig. 3.2 shows a simplified block diagram of the Cortex-M4 core architecture. Near the top left-hand side we can see the CPU which includes the ALU, register banks, instruction decoder, DSP module, and the memory interface. The DSP module is not available in the Cortex-M0 or Cortex-M3 cores. Notice that the memory is not part of the Cortex core and it is supplied by the company building the processor. The nested vectored interrupt controller is included at the top left corner of the figure. These are external interrupt inputs that can be used in interrupt-driven real-time applications. The core includes a FPU used to accelerate the floating point mathematical operations. The test and debug interface is shown at the bottom right-hand side of the figure. The bus matrix, memory protection unit, and peripheral interfaces are located at the bottom right hand-side of the figure. Notice that the core does not include the input-output ports.

Fig. 3.3 shows a microcontroller built around the Cortex core. Notice that this figure includes input-output ports, memory, and so on.

3.4 ARM PROCESSOR-BASED MICROCONTROLLER DEVELOPMENT BOARDS

There are many ARM processor-based development boards. In this section we are only interested in the Cortex-M-based development boards which are used in microcontroller-based

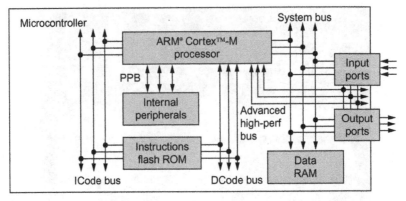

FIG. 3.3   Microcontroller built around Cortex core. *Used with permission of Arm Holdings.*

control and monitoring applications and which are also mbed compatible. Some popular development boards are described in this section. Details of mbed compatible ARM boards can be obtained from the following web site:

    https://os.mbed.com/platforms/

3.4.1 LPC11U24

This is a 32-bit Cortex-M0-based development board (Fig. 3.4) having the following features:

- 48 MHz clock
- 8 KB RAM, 32 KB flash memory
- 2 × SPI, I²C, UART, 6 × ADC, GPIO
- 40-pin 0.1″ DIP package
- built-in USB flash programmer

The development board is packaged in a small DIP form factor that can be used with a breadboard for quick development and testing.

3.4.2 LPC1768

This is a high-performance Cortex-M3-based development board (Fig. 3.5) having the following features:

- Cortex-M3 core
- 96 MHz clock
- 32 KB RAM, 512 KB flash memory
- Ethernet, 3 × SPI, 2 × I²C, 3 × UART, CAN, 6 × PWM, 6 × ADC, GPIO
- 40-pin 0.1″ DIP package
- 5 V USB supply
- Built-in USB flash programmer

FIG. 3.4 LPC11U24 development board.

FIG. 3.5 LPC1768 development board.

FIG. 3.6   EFM32 Zero Gecko starter kit.

This development board is similar to LPC11U24 but it is based on the high-performance Cortex-M3 core and includes more program and data memories.

3.4.3 EFM32 Zero Gecko Starter Kit

This is an ultralow-power 32-bit Cortex-M0+ core-based development board (Fig. 3.6). The board is ideal in applications requiring high performance and low-energy consumption. The basic features of this board are as follows:

- Cortex-M0+ core
- 24 MHz clock
- 4 KB RAM, 32 KB flash memory
- 37 GPIO
- 1.98–3.8 V power supply
- Ultralow current consumption (48 μA/MHz sleep mode)
- SPI, I^2C, 2×16-bit timers
- $2 \times$ user buttons, $2 \times$ LEDs
- $2 \times$ capacitive touch pads

3.4.4 Arch Pro

Arch Pro is based on a variant of the LPC1768 Cortex-M3 core. This development board (Fig. 3.7) has the following features:

- Cortex-M3 core
- 100 MHz clock
- 64 KB RAM, 512 KB flash memory
- $4 \times$ UART, $3 \times I^2C$, $2 \times$ SPI, Ethernet
- USB host/device
- Grove connectors
- Arduino form factor

FIG. 3.7   Arch Pro development board.

The advantage of this development board is that it supports Grove connectors and Arduino form factor and therefore large number of Grove modules and Arduino shields can be connected to the board.

3.4.5 LPC4088

This is a 44-pin development board based on the high-performance Cortex-M4 core (Fig. 3.8). The features of this board are as follows:

- Cortex-M4 core
- 120 MHz clock
- 8 MB QSPI + 512 KB on-chip, 32 MB SDRAM, 96 KB SRAM, 4 KB E2PROM
- 10/100 Mbps ethernet
- USB host/device interface
- XBee compatible connector
- 2 × 22 pin edge connector
- 4.5–5.5 V power supply, or power via micro-USB socket

FIG. 3.8   LPC4088 development board.

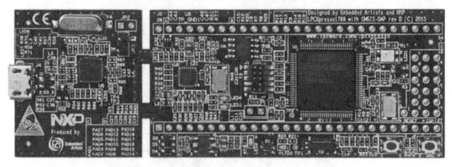

FIG. 3.9   LPC1769 LPCXpresso development board.

3.4.6 LPC1769 LPCXpresso

This development board (Fig. 3.9) is based on 32-bit Cortex-M3 core and has the following features:

- 120 MHz clock
- 4 MHz internal RC oscillator
- 64 KB RAM, 512 KB flash memory
- Ethernet
- USB device/host controller
- 8-channel DMA
- $4 \times$ UART, $2 \times$ CAN, $3 \times$ SPI, $3 \times I^2C$, I^2S, $8 \times$ ADC, DAC
- $4 \times$ timers, $6 \times$ PWM, RTC, 70 GPIO
- Quadrature encoder interface
- Power management unit
- Reduced power modes
- Code read protection
- On board SWD/JTAG debugger
- 3.15–3.3 V external power or power via USB port

3.4.7 LPC1347 LPCXpresso

This development board (Fig. 3.10) is based on 32-bit Cortex-M3 core and has the following features:

- Cortex-M3 core
- 72 MHz clock
- Low cost
- 12 KB SRAM, 64 KB flash memory
- ADC, $4 \times$ timers, $2 \times$ SPI, I^2C, USART
- USB bootloader

FIG. 3.10   LPC1347 LPCXpresso development board.

The LPCXpresso boards are actually two separate boards, and the debugger (called the LPC-Link) can actually be separated from the LPC1347 MCU half. This is useful if you plan to use the debugger with other boards inside the LPCXpresso IDE.

3.4.8 u-blox C027

This development board (Fig. 3.11) is based on Cortex-M3 core and has the following features:

- Cortex-M3 core
- 96 MHz clock
- 32 KB SRAM, 512 KB flash memory
- 6 × analog inputs
- 9 × PWM capable outputs
- 22 × GPIOs
- 1 × SPI
- 1 × I2C
- 1 × UART
- 1 × I2S
- Ethernet, CAN
- SIM card holder
- GNSS antenna SMA connector
- Cellular antenna SMA connector
- MAX-M8Q GPS/GNSS receiver
- LISA or SARA cellular module

3.4.9 Teensy 3.1

This development board (Fig. 3.12) is based on Cortex-M4 and has the following features:

- Cortex-M4 core
- 96 MHz clock
- 64 KB RAM, 256 KB flash, 2 KB EEPROM memory
- 34 × GPIO

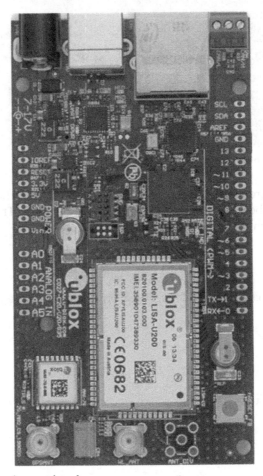

FIG. 3.11   u-blox C027 development board.

FIG. 3.12   Teensy 3.1 development board.

- 21 × analog inputs, 1 × analog output
- 2 × converters, 3 × comparators
- 12 × timers
- 3 × UART, SPI, 2 × I^{2}C, CAN, I^{2}S
- USB host
- RTC

3.4.10 DISCO-F334C8

This development board (Fig. 3.13) is based on Cortex-M4 core and has the following features:

- Cortex-M4 core
- 72 MHz clock

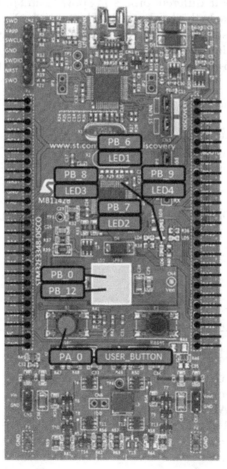

FIG. 3.13   DISCO-F334C8 development board.

- 16 KB SRAM, 64 KB flash memory
- 37 × GPIO
- 2 × analog inputs
- 3 × analog outputs
- 3 × USART, I^{2}C, SPI, CAN
- 9 × timers
- 2 × watchdog timers, 3 × analog comparators, operational amplifier
- On board debugger/programmer
- 4 × user LEDs, 2 × push-buttons

3.4.11 STM32 Nucleo Board Family

STM32 Nucleo boards are high-performance mbed compatible ARM core-based development boards manufactured by STMicroelectronics. The family includes large number of low-cost development boards with different processing powers and power consumption levels to suit all types of applications. In this book, we shall be using the Nucleo-F411RE development board in our projects. We shall be looking at the details of this development board in great detail in the next chapter. It is worthwhile to look at the features of this board here briefly. Fig. 3.14 shows the Nucleo-F411RE development board. Its basic features are:

- Cortex-M4 core
- STM32 microcontroller in LQFP64 package
- 1 user LED
- 1 user and 1 reset push-buttons
- 32.768 kHz LSE crystal oscillator
- Board expansion connectors:
  - Arduino Uno V3
  - ST morpho extension pin headers
- Flexible power-supply options: ST-LINK USB V_{BUS} or external sources
- On-board ST-LINK/V2-1 debugger/programmer
- Support for Mbed

3.5 SUMMARY

In this chapter we have learned about the following:

- History of the ARM processors
- CISC and RISC-type computers
- The architecture of the ARM processors
- Some popular Cortex-M development boards

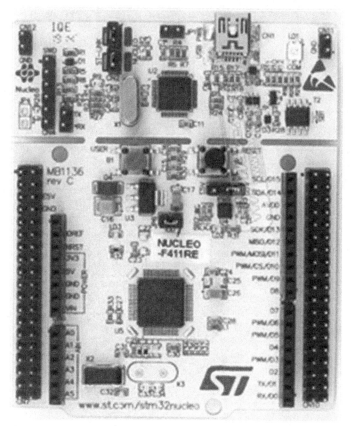

FIG. 3.14 Nucleo-F411RE development board.

3.6 EXERCISES

1. Describe the differences between the CISC- and RISC-type computers.
2. Which company gives licenses for the ARM core processors?
3. Explain the differences between Cortex-M and Cortex-A-type core processors.
4. What type of ARM core processors are commonly used in mobile phones?
5. Which Cortex processors have built-in DSP modules? Where can the DSP modules be used?
6. What type of ARM core processor will be used in the projects in this book?
7. Which ARM-based development board will be used in the projects in this book?

4

STM32 Nucleo Development Boards

4.1 OVERVIEW

This chapter is about the STM32 Nucleo development boards. Brief specifications of some of the Nucleo development boards are described in the chapter.

4.2 STM32 NUCLEO DEVELOPMENT BOARDS

The STMicroelectronics is an Italian-French multinational electronics manufacturing company with headquarters in Geneva, Switzerland. Also called ST for short, it is Europe's largest semiconductor chip maker. The company has 7400 people working in R&D, holds 17,000 patents, 9500 patent families, and has filed 500 new patents in 2017. The company employs 45,500 people and its revenue was $8.35 billion in 2017. With 80 sales and marketing offices in 35 countries and 11 main manufacturing sites, ST is one of the leading electronics technology innovator supplying high-quality microcontroller development boards, various electronics hardware, and software development tools.

The Nucleo family of low-cost development boards is small but powerful boards based on the state-of-the-art 32-bit ARM Cortex-M architecture. These development boards are targeted for a large audience, including students, professional engineers, and hobbyists at all levels. The boards are compatible with the popular Arduino, Mbed, ST-LINK, and ST Morpho, making it accessible to the users with different backgrounds. In addition to many hardware extension modules, the Nucleo family is supported by several software development tools and integrated development environments (IDEs), such as professional compilers, debuggers, and in-circuit programming tools.

There are over 30 different boards in the Nucleo family, aimed to satisfy the needs of almost all users with various backgrounds. Basically, the Nucleo boards come in three different sizes where the numbers below refer to the pin counts of the MCUs used on the boards:

- Small (Nucleo-32)
- Short (Nucleo-64)
- Long (Nucleo-144)

These three groups are further divided into three subgroups, identified by three different colors:

- ultralow power (green)
- mainstream (blue)
- high performance (magenta)

The ultralow-power boards are based on the STM32 L family and these boards are targeted for low-power applications, such as watches, smart meters, etc. Examples of the ultralow-power boards are: Nucleo-L011K4, Nucleo-L031K6, Nucleo-L432KC, and Nucleo-L433RC-P. There are three subcategories in the STM32 L family:

- L0, ARM Cortex-M0+
- L1, ARM Cortex-M3
- L4, ARM Cortex-M4

About half of the STM32 Nucleo boards are in the mainstream category. Examples of the mainstream boards are: Nucleo-F303K8, Nucleo-F042K6, Nucleo-F303RE, etc. There are three subcategories in the mainstream category:

- F0, ARM Cortex-M0+
- F1, ARM Cortex-M3
- F3, ARM Cortex-M4

The high-performance boards have large memories and faster MCUs. Examples of high-performance boards are: Nucleo-F410RB, Nucleo-F401RE, Nucleo-F722ZE, etc. There are three subcategories in the high-performance category:

- F2, ARM Cortex-M3
- F4, ARM Cortex-M4
- F7, ARM Cortex-M7

The Nucleo-32 boards are small (50 mm × 19 mm) and are Arduino Nano compatible. The Nucleo-64 and Nucleo-144 boards are Arduino Uno compatible and they also have the standard ST Morpho extension connectors which carry the MCU pins. There are a large number of Arduino Nano/Uno compatible shields available in the market and these shields can easily be used with the Nucleo boards, thus making it easy to quickly develop projects using the Nucleo boards.

Depending upon the model, the Nucleo boards have flash program memory sizes ranging from 16 KB to 2 MB, and RAM memories ranging from 4 to 320 KB. The clock frequency varies from 32 to 216 MHz.

Fig. 4.1 shows a comparison of the Nucleo boards available at the time this book was written.

4.2.1 Nucleo-32 Development Boards

Fig. 4.2 shows an example of Nucleo-32 board, the Nucleo-L031K6. This is an ultralow-power low-cost board incorporating the 32-pin STM32L031K6T6 microcontroller. The board

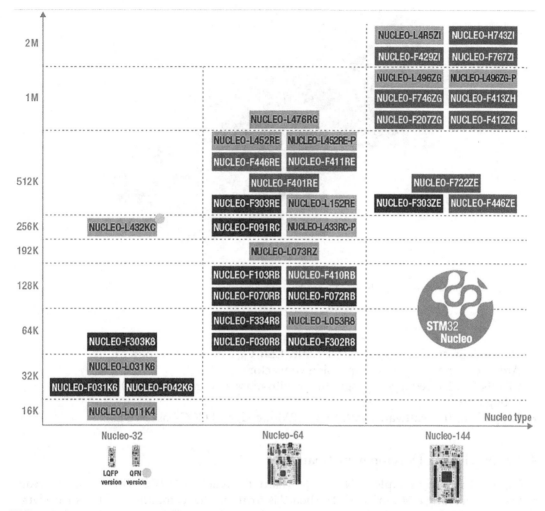

FIG. 4.1   Nucleo boards. *(Used with permission from ©STMicroelectronics.)*

is Arduino Nano compatible so that a large number of Arduino Nano shields can be used with the board. This board has the following features:

- 32 MHz Cortex M0+ microcontroller in 32-pin package
- 32 KB flash memory
- 8 KB RAM
- 1 KB EEPROM
- Real-time clock
- Serial interfaces (USART, SPI, and I2C)
- 3 LEDs (USB communication, power, and user)
- Push-button Reset

FIG. 4.2   Nucleo-32 development board (Nucleo-L031K6).

- Flexible power-supply options: ST-LINK USB V_{BUS} or external sources
- Arduino Nano compatible expansion connector
- ST-LINK/V2-1 debugger/programmer with mass storage, virtual COM port, and debug port
- Support for IDE software (IAR, Keil, ARM mbed, and GCC-based IDEs).

4.2.2 Nucleo-64 Development Boards

Fig. 4.3 shows an example of Nucleo-64 board, the Nucleo-F091RC. This is a mainstream board incorporating a 64-pin MCU. The board is Arduino Uno compatible and as such a large number of Arduino Uno shields can be used with the board. This board has the following features:

- 1 user LED
- 1 user push-button switch
- 32.768 kHz crystal oscillator
- ST morpho connector
- Arduino Uno expansion socket
- Flexible power-supply options: ST-LINK USB V_{BUS} or external sources
- ST-LINK/V2-1 debugger/programmer with mass storage, virtual COM port, and debug port
- Comprehensive free software libraries

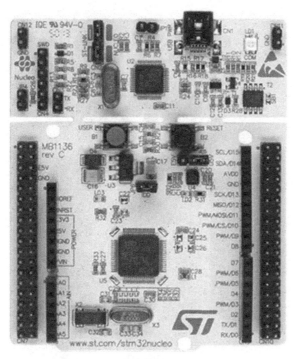

FIG. 4.3 Nucleo-64 development board (Nucleo-F091RC).

- Support of a wide choice of IDE software (IAR, Keil, ARM mbed, and GCC-based IDEs)

4.2.3 Nucleo-144 Development Boards

Fig. 4.4 shows an example of Nucleo-144 board, the Nucleo-F722ZE. This is a high-performance board incorporating a 144-pin MCU. The board is Arduino Uno compatible. The features of this board are:

- Ethernet compliant with RJ45 connector
- ST morpho connector
- ST-LINK/V2-1 debugger/programmer with mass storage, virtual COM port, and debug port
- ST Zio connector
- 3 user LEDS
- 2 push-button switches
- 32.768 kHz crystal oscillator
- Flexible power-supply options: ST-LINK USB V_{BUS} or external sources
- Comprehensive free software libraries

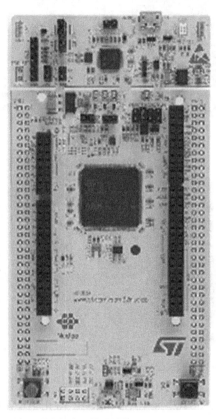

FIG. 4.4   Nucleo-144 development board (Nucleo-F7222ZE).

- Support of a wide choice of IDE software (IAR, Keil, ARM mbed, and GCC-based IDEs)

4.3 STM32 NUCLEO EXPANSION BOARDS

A large number of expansion boards are offered by the STMicroelectronics. These expansion boards are fully compatible with the Nucleo development boards and they just plug on top of the development boards. There are many expansion boards that can be used in many sensor and actuator-based applications. Some of the commonly used expansion board provide the following functionalities:

- Wi-Fi
- Bluetooth
- Brushed DC motor controller
- Stepper motor controller
- Brushless DC motor controller

- LED driver
- NFC card detection
- Industrial input-output
- Microelectromechanical system (MEMS) accelerometer, gyroscope, temperature sensor, humidity sensor, and atmospheric pressure sensor
- Relay output board

For example, the Wi-Fi expansion board can be used when it is required to develop a microcontroller-based application using a Nucleo development board to establish Wi-Fi-based communication (e.g., an Internet application, or remote monitoring and control application). All the user has to do is plug the Wi-Fi expansion board on top of the Nucleo development board and power up the Nucleo board. The required software can then be developed easily using, for example, the Wi-Fi library provided by Mbed. In addition to the libraries, various sample programs are provided by Mbed for using the expansion boards in projects. With the help of these libraries, a project can be developed in considerably less time since the expansion board has already been tested and working. It is also an advantage that the user does not need to know much about the configuration or the hardware design of the expansion board.

4.4 SUMMARY

In this chapter, we have learned about the following:

- Nucleo board family
- Types of Nucleo boards
- Basic features of the various Nucleo boards
- Nucleo expansion boards

4.5 EXERCISES

1. Explain how many types of Nucleo boards are there.
2. Draw a table and compare the features of different types of Nucleo boards.

The Nucleo-F411RE Development Board

5.1 OVERVIEW

This chapter is about the Nucleo-F411RE development board which is the board used in the projects in this book. The chapter describes the structure of this development board in great detail.

5.2 THE NUCLEO-F411RE DEVELOPMENT BOARD

The Nucleo-F411RE board is one of the most popular Nucleo development boards available at the time of writing this book. This is a 64-pin low-power board (see Fig. 5.1) based on the Cortex-M4 core and uses the STM32F411RET6 microcontroller developed by the STMicroelectronics. This board will be used as the example development board in all the projects in this book. Therefore, full details of this board are given in this section so that the readers are familiar with this board. Further details about the Nucleo-F411RE board can be obtained from the STMicroelectronics **UM1724 User Manual (DocID025833 rev 12)**.

Notice that in the model number Nucleo-F411RE, **R** refers to 64-pin MCU and **E** refers to 512 kB flash memory.

5.2.1 The Board

The Nucleo-F411RE board measures 70 mm × 82.5 mm. As shown in Fig. 5.2, the board consists of two parts: the smaller ST-LINK part and the STM32 MCU part. The MCU part contains the MCU, two push-buttons, LEDs, Arduino and ST morpho connectors, the crystal, and the power controller. The ST-LINK part contains the micro-USB port and the programming/debugging interface. The ST-LINK part of the printed circuit board (PCB) can be cut if desired to reduce the overall board size. If this is done, the MCU part can only be powered by VIN,

FIG. 5.1   Nucleo-F411RE development board.

E5V, and 3.3 V on the ST Morpho connector CN7, or by VIN and 3.3 V Arduino connector CN6. It is possible to program the MCU after the ST-LINK part is cut by connecting wires between CN4 on the ST-LINK board and serial wire debug (SWD) signals available on the ST Morpho connector (CN7 pin 15 SWCLK and CN7 pin 13 SWDIO).

The components on the top side of the board are shown in Fig. 5.3. Some of the important components that you may need to know their locations on the board are (starting from the top left-hand side of the board):

CN1: mini USB socket
CN2: ST-LINK/Nucleo selector
CN4: SWD connector
B1: User push-button

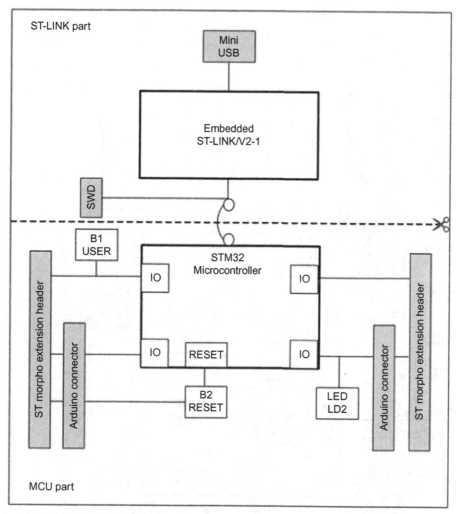

FIG. 5.2 Nucleo-L433RC-P board parts. *(Used with permission from ©STMicroelecronics.)*

JP6: Current measurement
LD3: +5 V Power LED
CN6: Arduino connector
CN7: ST morpho connector
CN8: Arduino connector
CN9: Arduino connector
CN10: ST morpho connector
CN5: Arduino connector
LD2: Green LED
JP5: Power selection

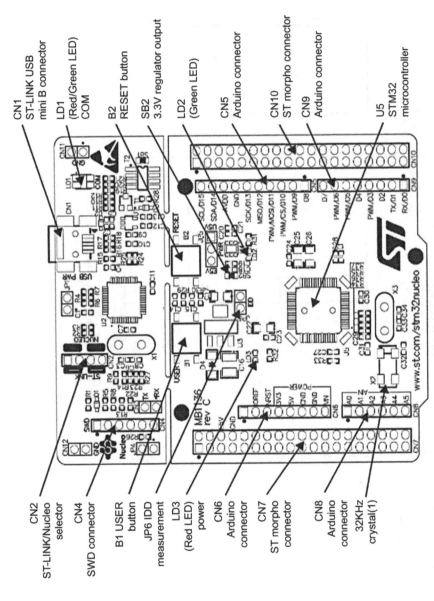

FIG. 5.3 Components on the top side of the board. (*Used with permission from ©STMicroelectronics.*)

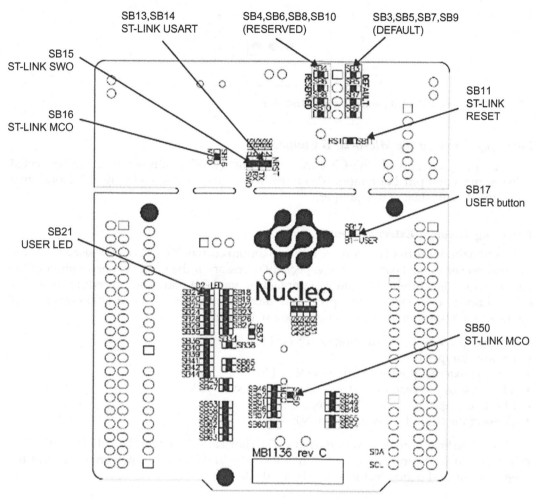

FIG. 5.4 Bottom side of the board. *(Used with permission from ©STMicroelecronics.)*

B2: Reset button
LD1: Red/Green communication LED

The bottom side of board is shown in Fig. 5.4.

5.2.2 The Power Supply

DC power to the board can be supplied using one of the following sources: 5 V from the USB connector, VIN (7–12 V), E5V (5 V), or +3.3 V from power-supply pins on connectors CN6 or CN7.

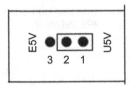

FIG. 5.5   Powering the board through the mini-USB port.

Powering Through the Micro-USB Connector

When powered through the CN1 mini-USB port (U5V), a jumper must be connected between pin 1 and pin 2 of jumper JP5 on the MCU board as shown in Fig. 5.5. Notice that this is the default state of this jumper.

Powering Through External Inputs

VIN (maximum current 800 mA) or E5V (maximum current 500 mA) can be used to supply external power to the board. This is especially necessary if the current consumption of the board exceed the current that can be supplied by the USB port. The procedure to supply external power is as follows (this procedure must be adhered to so that USB power is not applied to the board at the same time as external power):

- Connect pin 2 and pin 3 of jumper JP5 (see Fig. 5.6)
- Remove jumper JP1
- Connect external power source to VIN or E5V
- VIN must be between +7 and +12 V, or E5V must be +5 V
- Check to make sure that LD3 is ON
- Connect the PC to USB connector CN1

+3.3 V: The board can be powered from +3.3 V (pin 4 of CN6, or pin 12 and pin 16 of CN7). When the Nucleo board is powered from +3.3 V, the ST-LINK is not powered and thus the programming and debug features are not available.

Power Output

In some applications, we may want to power external devices. +5 V voltage output is available from pin 5 of connector CN6 or pin 18 of connector CN7 when the board is powered by USB, VIN, or E5V.

We can also get +3.3 V output from pin 4 of connector CN6 or pins 12 and 16 of connector CN7.

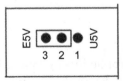

FIG. 5.6   Powering the board from VIN or E5V.

5.2.3 The LEDs

There are two LEDs on the board. The functions of these LEDs are as follows:

LD1: This is a three-color LED (green/orange/red) with default color of red which turns green to indicate that communication is taking place between the PC and the board.
LD2: This is the green color user LED connected to Arduino signal D13 (port PA_5 of Nucleo-F411RE). The LED turns ON when logic 1 is applied to the corresponding I/O pin, and turns OFF when logic 0 is applied to the pin. The user LED is also known as LED1.

5.2.4 Push-Button Switches

There are two push-button switches on the MCU board called B1 and B2.

B1 is the user button and is connected to I/O pin PC13 of the Nucleo-F411RE board. The user button is also known as BUTTON1.
B2 is the Reset button used to reset the MCU and is connected to NRST. The button state is normally at logic 1 and goes to logic 0 when pressed.

5.2.5 Current Measurement

This jumper JP6 is labeled IDD and can be used to measure the current consumption of the MCU by removing the jumper and then connecting an ammeter to the jumper pins. This jumper is ON by default, that is, the board is shipped with the current measurement function disabled.

5.2.6 The ST-LINK/V2-1

The ST-LINKV2-1 programming and debugging tool is integrated in the Nucleo boards and it makes the boards mbed enabled. The ST-LINK/V2-1 supports only SWD for the STM32 devices. The ST-LINK/V2-1 does not support single wire interface module (SWIM) interface and the minimum supported application voltage is limited to 3 V. The ST-LINK/V2-1 supports virtual COM port interface on USB, USB software re-numeration, mass storage interface on USB, and USB power management request for more than 100 mA power on USB.

In order to program the board, we have to plug in two jumpers on connector CN2 as shown in Fig. 5.7. Table 5.1 shows the connector CN4 configurations.

Before connecting the Nucleo-64 board to a Windows PC, a driver for the ST-LINK/V2-1 must be installed. This can be downloaded from the following site. You will have to register at the site so that you can download the driver. At the time of writing this book, the driver was called **en.stsw-link009.zip**:

http://www.st.com/en/development-tools/stsw-link009.html#getsoftware-scroll

5.2.7 Input-Output Connectors

Fig. 5.8 shows the input-output connectors for the Nucleo-F411RE board. Notice that different models may have different pin configurations and you should check the appropriate

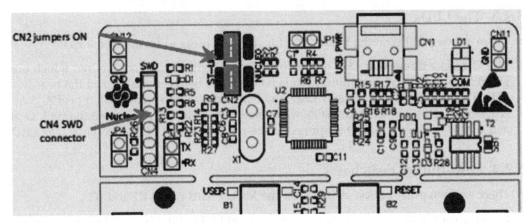

FIG. 5.7   Connector CN2.

TABLE 5.1   Connector CN4 Configurations

Jumper	Function	Default State	Description
JP1	ST-LINK RST	ON[1-2]	Reset MCU
		OFF	**Normal use**
JP2/JP3	Ground	**OFF**	**Ground probe**
JP4	nRST	**ON**	**ST-LINK can reset MCU**
		OFF	ST-LINK cannot reset MCU
JP5	5 V Power selection	**ON[1-2]**	**5 V from ST-LINK**
		ON[3-4]	5 V from VIN
		ON[5-6]	5 V from E5V
		ON[7-8]	5 V from USB_CHARGE
		OFF	No 5 V power (use 3.3 V)
JP6	Current measurement	**ON[1-2]**	**No current measurement**
		OFF	Current measurement mode
JP7	VDD_MCU = 3.3 V	**ON[1-2]**	**VDD_MCU voltage = 3.3 V**
		ON[2-3]	VDD_MCU voltage = 1.8 V
		OFF	No VDD_MCU
JP8	VDD_IN_SMPS	**ON[1-2]**	**1.1 V ext SMPS input power**
		OFF	Ext SMPS not powered
CN2		**ON[1-2], ON[3-4]**	**ST-LINK enable for debugger**
		OFF	ST-LINK enabled for ext CN2 connector

User Manual for the correct configuration of the model you are using. CN7, CN8, CN9, and CN10 are Arduino UNO compatible connectors. CN5 and CN6 are the ST Morpho connectors that carry most of the MCU signals. Jumper wires can be used to make connections to the external sensors or devices through these input-output connectors. The Nucleo expansion

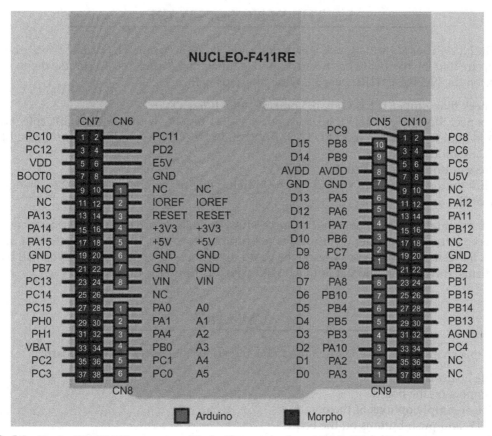

FIG. 5.8 Nucleo-F411RE I/O connectors. (*Used with permission from ©STMicroelecronics.*)

boards are usually connected to the Nucleo development boards using the Arduino UNO compatible connectors.

5.2.8 Jumpers at the Bottom of the Board

A number of tiny jumpers (or solder tags) are provided at the bottom of the board for configuring the board. These jumpers can be de-soldered or soldered as required by the application. In normal operations, it will not be necessary to change any of these jumper positions. These jumpers may need to be changed if, for example, any of the following are required (interested readers should refer to the **UM1724 User Manual** for more information on these jumpers):

- Changing the default USB UART port (USBTX and USBRX)
- Changing the default user LED port
- Changing the default user push-button port

5.3 THE DEMO SOFTWARE

The Nucleo boards are shipped with a preloaded demo software. The demo software helps to ensure that the hardware has no errors and is functioning correctly. To run the demo software on the Nucleo-F411RE board, follow the steps below:

- Check that jumper JP1 is OFF, JP5 on U5V side, JP6 is ON.
- Connect the Nucleo-F411RE board connector CN1 to a PC with a micro-USB cable (type A to mini-B cable). The red LED LD3 (PWR) and LD1 (COM) should light up. Green LED LD2 should blink.
- Press button B1 (left button).
- Click button B1 and you should see the blinking rate of the green LED LD2 to change accordingly.

If the demonstration software is run successfully, then the board should be ready for use in projects. If the demonstration software is not running correctly, then you should first of all check the default jumper positions on the board, then check your USB cable. If you still have problems, then you should get in touch with your distributor.

5.4 SUMMARY

In this chapter, we have learned the following about the Nucleo-F411RE board:

- The board layout
- Jumpers on the board
- Power-supply options of the board
- LEDs and push-buttons on the board
- Pin configuration of the board
- Operating the board in demo mode

5.5 EXERCISES

1. Explain how the Nucleo-F411RE development board can be powered from external power sources.
2. How many LEDs are there on the Nucleo-F411RE board? Explain the functions of these LEDs?
3. How many push-buttons are there on the Nucleo-F411RE board? Explain the functions of these buttons.
4. How many edge connectors are there on the Nucleo-F411RE board? What are their names?
5. Explain the functions of the various jumpers on the board.
6. Explain how the board can be operated in demo mode.

6

Architecture of the STM32F411RET6 Microcontroller

6.1 OVERVIEW

In this book, we shall be using the STM32 Nucleo-F411RE development board. This board incorporates the 32-bit, 64-pin STM32F411RET6 microcontroller. In this chapter, we shall be looking at the features of the STM32F411RET6 microcontroller. It is important to know the basic internal architecture of a microcontroller before it is used successfully in a project. The internal architecture of this microcontroller is very complex and we shall only briefly look at the commonly used features of this microcontroller, such as the GPIO, timers, analog-to-digital converter (ADC) and digital-to-analog converter (DAC), interrupt controller, power management, and so on. Interested readers can get detailed information on this microcontroller from the following STMicroelectronics website:

https://www.st.com/en/microcontrollers/stm32l433rc.html

6.2 KEY FEATURES OF THE STM32F411RET6 MICROCONTROLLER

The STM32F411RET6 microcontroller is based on the Cortex-M4 core and it is used on the Nucleo-F411RE development board. The key features of this microcontroller are as follows (see the above website):

- Cortex-M4 core
- 1.7–3.6 V power supply
- Frequency up to 100 MHz
- 512 KB flash memory
- 128 KB static random-access memory (SRAM)
- Internal 16 MHz RC oscillator
- 4–26 MHz crystal oscillator
- Internal 32 kHz RC oscillator

- 100 µA/MHz run time power consumption
- 16-Stream direct memory access (DMA) controllers
- 11 Timers
- Serial wire debug (SWD) and JTAG interface
- 52 I/O ports
- 3 × I2C interface
- 3 × USART
- 5 × SPI/I2S
- SD/MMC/eMMC interface
- Real-time clock (RTC) with hardware calendar
- Cyclic redundancy check (CRC) calculation unit

Fig. 6.1 shows the block diagram of the STM32F411RET6 microcontroller, showing all the modules supported by the microcontroller. The Cortex-M4 core is shown at the top middle part of the figure. The I/O connectivity modules are shown at the top left-hand part of the figure. Timer modules are shown at the middle right-hand side of the figure. The GPIO and the analog input and output modules are shown at the bottom left and bottom right-hand side of the figure, respectively.

Fig. 6.2 shows the pin configuration of the STM32F411RET6 microcontroller. The pin definitions are as follows (some pins have more than one function):

The internal structure of the STM32F411RET6 microcontroller is shown in Fig. 6.3A and B.

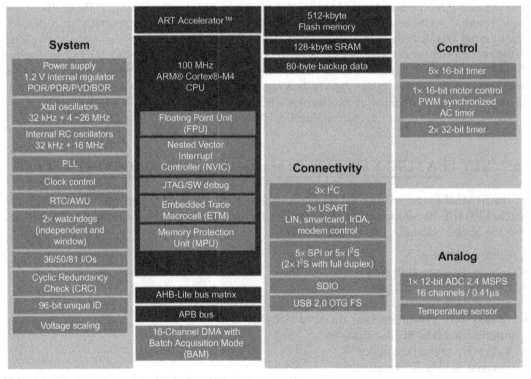

FIG. 6.1 Block diagram of the STM32F411RET6 microcontroller.

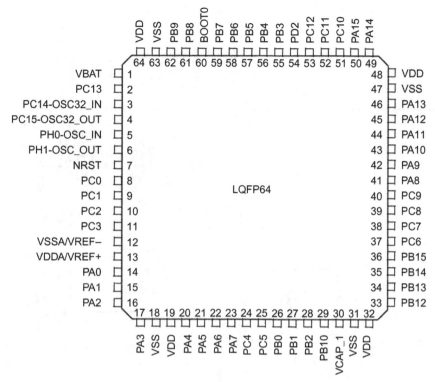

FIG. 6.2 Pin configuration of the microcontroller.

PA0–PA15	GPIO Port A pins
PB0–PB15	GPIO Port B pins
PC0–PC15	GPIO Port C pins
PD2	GPIO Port D pin
PH0–PH1	GPIO Port H pins
OSC_IN, OSC_OUT	Main crystal oscillator pins
OSC32_IN, OSC32_OUT	32 kHz crystal oscillator pins
VDD, VSS	Power and ground pins
VSSA, VDDA	Reference voltage pins
NRST	Reset pin
VBAT	External battery pin
BOOT0	Boot0 pin
VDD12	External SMPS power pin

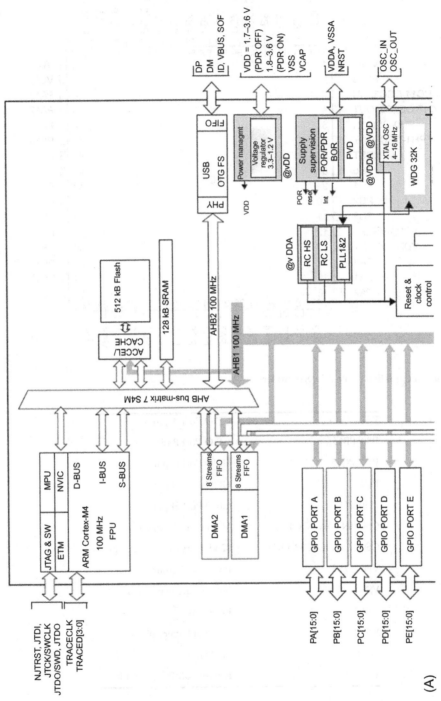

(A)

FIG. 6.3 (A) and (B) Internal structure of the STM32F411RET6 microcontroller.

(Continued)

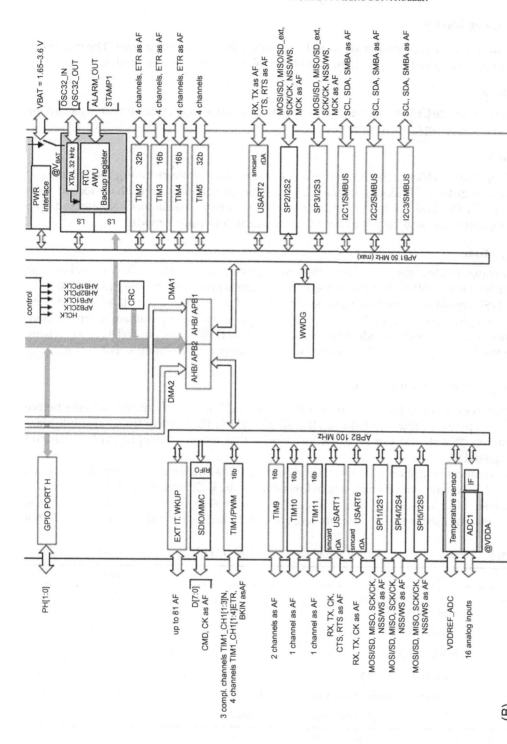

(B)

FIG. 6.3, CONT'D

6.2.1 Power Modes

By default, the microcontroller starts in Run mode after power-up or reset. The microcontroller can be configured to operate in one of the following low-power modes.

Sleep Mode

In Sleep mode, the CPU is stopped, all peripherals continue to operate and can wake up the CPU when an interrupt/event occurs.

Stop Mode

The Stop mode achieves the lowest power consumption while retaining the contents of SRAM and registers. All clocks in the 1.2 V domain are stopped, the phase-locked loop (PLL), the HSI RC, and the high-speed external (HSE) crystal oscillators are disabled. The voltage regulator can also be put either in normal or in low-power mode.

Standby Mode

The Standby mode is used to achieve the lowest power consumption. The internal voltage regulator is switched off so that the entire 1.2 V domain is powered off. The PLL, the HSI RC, and the HSE crystal oscillators are also switched off. After entering the Standby mode, the SRAM and register contents are lost except for registers in the backup domain when selected. The devices exit the Standby mode when an external reset (NRST pin), an IWDG reset, a rising edge on the WKUP pin, or an RTC alarm/wakeup/tamper/time stamp event occurs.

6.2.2 Electrical Characteristics

It is important to know the absolute maximum electrical ratings of a microcontroller before it is used in a project. Table 6.1 illustrates the STM32F411RET6 microcontroller absolute maximum ratings that should not be exceeded. Notice that the power-supply voltage should never exceed 4.0 V.

The current consumption characteristic of the STM32F411RET6 microcontroller is shown in Table 6.2. Two important parameters in this table are the I/O current capability of each I/O pin which should not exceed 20 mA. This current is normally sufficient to drive small LEDs, switches, etc. Transistor switch circuits (e.g., MOSFET switch) should be used if external

TABLE 6.1   Absolute Maximum Electrical Ratings of the STM32F411RET6 Microcontroller

Symbol	Ratings	Min	Max	Unit
$V_{DD} - V_{SS}$	External main supply voltage (including V_{DDA}, V_{DD}, and V_{BAT})[a]	−0.3	4.0	V
V_{IN}	Input voltage on FT and TC pins[b]	V_{SS} −0.3	V_{DD} +4.0	
	Input voltage on any other pin	V_{SS} −0.3	4.0	
	Input voltage for BOOT0	V_{SS}	9.0	
$|\Delta V_{DDx}|$	Variations between different V_{DD} power pins	–	50	mV
$|V_{SSx} - V_{SS}|$	Variations between all the different ground pins	–	50	

[a]*All main power and ground pins must always be connected to the external power supply.*
[b]*V_{IN} maximum value must always be respected.*

TABLE 6.2 Current Characteristics of the STM32F411RET6 Microcontroller

Symbol	Ratings	Max.	Unit
ΣI_{VDD}	Total current into sum of all V_{DD_x} power lines (source)[a]	160	mA
ΣI_{VSS}	Total current out of sum of all V_{SS_x} ground lines (sink)[a]	−160	
I_{VDD}	Maximum current into each V_{DD_x} power line (source)[a]	100	
I_{VSS}	Maximum current out of each V_{SS_x} ground line (sink)[a]	−100	
I_{IO}	Output current sunk by any I/O and control pin	25	
	Output current sourced by any I/O and control pin	−25	
ΣI_{IO}	Total output current sunk by sum of all I/O and control pins[b]	120	
	Total output current sourced by sum of all l/Os and control pins[b]	−120	
$I_{INJ(PIN)}$[c]	Injected current on FT and TC pins[d]	−5/+0	
	Injected current on NRST and B pins[d]		
$\Sigma I_{INJ(PIN)}$	Total injected current (sum of all I/O and control pins)[e]	±25	

[a]All main power and ground pins must always be connected to the external power supply.
[b]This current consumption must correct be distributed over all I/O and control pins.
[c]Negative injection disturbs the analog performance of the device.
[d]Positive injection is not possible on these pins.
[e]When several inputs are submitted to current injection, the maximum is the absolute sum of the positive and negative injected currents.

devices requiring more than 20 mA are connected to an I/O pin. Also, the total current of all the I/O pins should not exceed 100 mA.

6.2.3 The Interrupt Controller

A nested vectored interrupt controller is available on the microcontroller which can manage 16 interrupt priority levels, and can handle up to 62 maskable interrupt channels with 16 interrupt lines of the Cortex-M4. Interrupt management is provided with minimal interrupt latency which allows early processing of interrupts. The processor state is automatically saved on entry to an interrupt, and restored on interrupt exit, with no instruction overhead.

6.2.4 The Analog-to-Digital Converter (ADC)

A successive approximation ADC is used with the following specifications:

- 12-bit resolution
- 5.33 Msps conversion rate (higher for lower resolution)
- 18.75 ns sampling time
- Up to 16 channels
- Internal reference voltage
- Single-ended and differential mode inputs
- Dual clock operation
- Single-shot or continuous mode
- Multiple trigger inputs

- Results stored in data register (or RAM)
- Data preprocessing (left/right alignment) and offset compensation
- Built-in oversampling unit
- Programmable sampling time

6.2.5 Digital-to-Analog Converter (DAC)

Two 12-bit DAC channels are available on the chip. The DAC has the following features:

- 8-bit or 12-bit output mode
- Left/right data alignment (in 12-bit mode)
- Noise-wave generation
- Triangular-wave generation
- Dual DAC independent or simultaneous conversion
- DMA capability
- External triggering
- Sample and hold with internal or external capacitor

6.2.6 Temperature Sensor

A temperature sensor is available on the microcontroller chip that generates voltage linearly proportional with temperature. This sensor is internally connected to the ADC_IN18 analog input channel of the microcontroller. The sensor should be calibrated for good accuracy as its characteristics change from chip to chip due to process variations. The devices are calibrated at factory by ST where the calibration data are available in the system memory as read-only data (see the device data sheet).

6.2.7 Timers and Watchdogs

Timer modules are important parts of all microcontrollers. The STM32F411RET6 microcontroller includes one advanced control timer, seven general purpose timers, two watchdog timers, and a SysTick timer. Table 6.3 presents a brief summary of the timers.

6.2.8 The Clock Circuit

The clock circuit is an important part of the microcontroller. As shown in Fig. 6.4, there two system clock (SYSCLK) sources: *External* clock sources and *Internal* clock sources.

External Clock Sources

High Speed External (HSE): This can be an external crystal or resonator device, or an external clock signal. The frequency range of the crystal or resonator should be 4–48 MHz. It is recommended to use two capacitors in the range of 4–25 pF with the crystal circuit.

When using a clock generator circuit, the waveform can be square, sine, or triangular, and the waveform must be symmetrical, that is, 50% ON and 50% OFF times. The clock signal must be fed to the OSC_IN pin of the microcontroller. If external clock circuitry is used, then HSE oscillator should be bypassed to avoid any conflict.

TABLE 6.3 The STM32F411RET6 Microcontroller Timers

Timer Type	Timer	Counter Resolution	Counter Type	Prescaler Factor	DMA Request Generation	Capture/ Compare Channels	Complementary Output	Max. Interface Clock (MHz)	Max. Timer Clock (MHz)
Advanced-control	TIM1	16-bit	Up, down, up/down	Any integer between 1 and 65536	Yes	4	Yes	100	100
General purpose	TIM2, TIM5	32-bit	Up, down, up/down	Any integer between 1 and 65536	Yes	4	No	50	100
	TIM3, TIM4	16-bit	Up, down, up/down	Any integer between 1 and 65536	Yes	4	No	50	100
	TIM9	16-bit	Up	Any integer between 1 and 65536	No	2	No	100	100
	TIM10, TIM11	16-bit	Up	Any integer between 1 and 65536	No	1	No	100	100

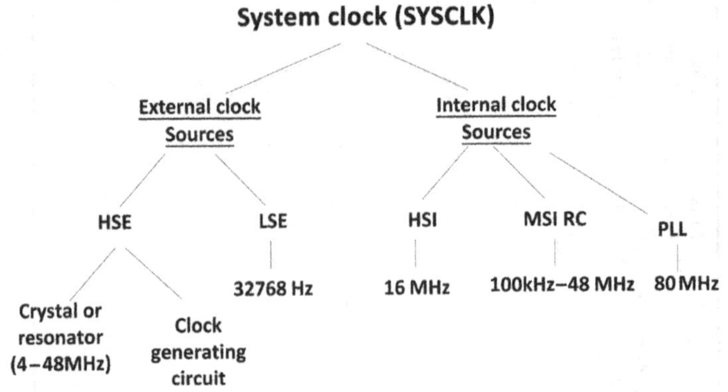

System clock (SYSCLK)

FIG. 6.4   System clock sources.

Low-Speed External (LSE): This is a 32,768 Hz clock driven from an external crystal and feeding the internal RTC module.

Internal Clock Sources

High-Speed Internal (HSI): This is an accurate RC-based 16 MHz internal clock with a factory calibrated tolerance of 1%.

Multispeed Internal (MSI RC): This is a multispeed RC-based clock source providing clock in the range 100 kHz–48 MHz. This clock can be trimmed by software and is able to generate 12 different clock frequencies.

PLL: The PLL is fed by HSE, HSI16, or MSI clocks and it can generate system clocks up to 100 MHz.

Microcontroller Clock Output (MCO)

A clock output is possible from a special pin called *MCO*. This clock output can be used as a general purpose clock or as a clock for another microcontroller.

Low-Speed Clock Output (LSCO)

A low-speed (32,768 kHz) clock (LSCO) is available as a general purpose clock output.

Several prescalers are available in the clock circuit to configure the AHB, APB1, and APB2 bus clock frequencies for the system and the peripheral devices. The maximum frequency of the AHB and APBx is 80 MHz.

Configuring the Clock

As shown in Figs. 6.5 and 6.6, the clock circuit consists of a number of multiplexers, prescalers, and a PLL. The multiplexers are used to select the required clock source. The prescalers are used to divide the clock frequency by a constant. Similarly, the PLL is used to multiply the clock frequency with a constant in order to operate the chip at higher frequencies.

It is important to select the correct clock source for an application. Configuring the clock sources by programming the internal clock registers is a complex task and detailed

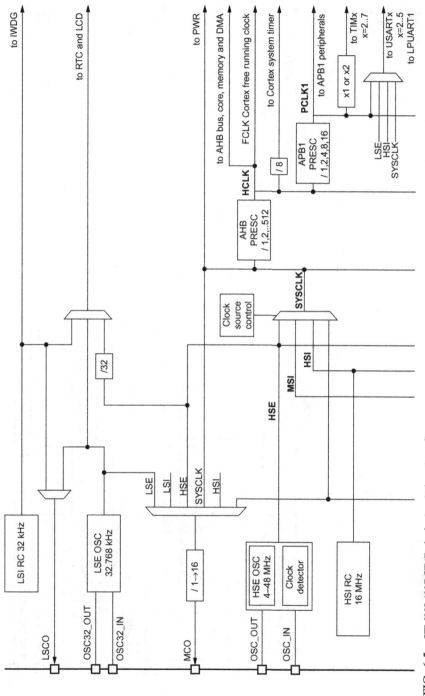

FIG. 6.5 STM32L433RCT6P clock circuit (continued).

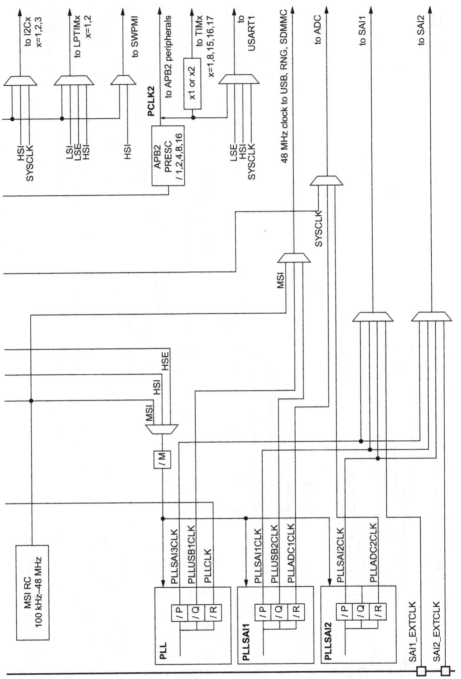

FIG. 6.6   STM32L433RCT6P clock circuit.

knowledge of the clock circuitry is required. In reference to Figs. 6.5 and 6.6, we can identify the following prescalers:

AHB prescaler: this prescaler receives the system clock SYSCLK and provides clock to the AHB bus (HCLK). The prescaler can take the values from 1,2 to up to 512.
APB1 prescaler: this prescaler receives the AHB clock (HCLK) and provides clock to the APB1 bus peripherals. The prescaler can take the values 1,2,4,8, and 16.
APB2 prescaler: this prescaler receives the AHB clock (HCLK) and provides clock to the APB2 bus peripherals. The prescaler can take the values 1,2,4,8, and 16.

6.2.9 General Purpose Input and Output (GPIO)

The STM32F411RET6 microcontroller has 64 pins and 52 of them can be used as general purpose input or output (GPIO). The GPIO is arranged into five ports, where ports A, B, and C are 16-bits wide, port H is 3-bits wide, and port D is only 1-bit. During and after reset, most ports are configured in analog input mode. Each port has the following basic features:

- all GPIO pins have weak internal pull-up and pull-down resistors, which can be activated under software control.
- port input states can be floating, pull-up, pull-down, or analog.
- port outputs can be configured as push-pull, open-drain or pull-up, pull-down
- the speed of each port can be set
- I/O port configurations can be locked under software control
- port pins can be digital I/O, analog input, or they can have alternative functions, such as DAC, SPI, USB, PWM, etc.)
- each port pin can be used with one of 15 alternate functions (AFs)
- bit manipulations can be performed on each port pin

Each GPIO port, subject to its hardware characteristics, can be configured in software in several modes:

- input pull-up
- input-pull-down
- analog
- output open-drain with pull-up or pull-down capability
- Output push-pull with pull-up or pull-down capability
- AF push-pull with pull-up or pull-down capability
- AF open-drain with pull-up or pull-down capability

Fig. 6.7 shows the structure of a push-pull output port pin. Similarly, an open-drain output port pin is shown in Fig. 6.8. Input pull-up and pull-down circuits are shown in Figs. 6.9 and 6.10, respectively.

GPIO port pins can be programmed for AFs. For AF inputs, the port must be configured in the required input mode. Similarly, for AF outputs, the port must be configured in AF output mode. When a port is configured as AF, the pull-up and pull-down resistors are disabled, the output is set to operate in push-pull or in open-drain mode. When an I/O port is configured as AF:

- the output buffer can be configured in open-drain or push-pull mode
- the output buffer is driven by the signals coming from the peripheral

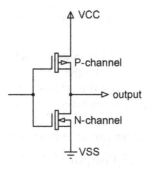

FIG. 6.7 Push-pull output pin.

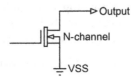

FIG. 6.8 Open-drain output pin.

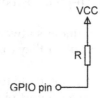

FIG. 6.9 Pull-up pin.

FIG. 6.10 Pull-down pin.

- the Schmitt trigger input is activated
- the data present on the I/O pin are readapt every AHB clock cycle

The GPIO pins are +5 V tolerant (except pins PA0 and PB5). +5 V tolerant GPIO pins are denoted with letters FT in the data sheet.

6.2.10 I^{2}C Interface

The microcontroller supports three I^{2}C bus interfaces that can operate in multimaster and in slave modes. The I^{2}C interface can operate at up to 400 kHz, but it can be increased up to 1 MHz. A hardware CRC generation/verification is included.

6.2.11 SPI Interface

Five SPI interfaces in slave and master modes are supported. Three SPI interfaces can operate at up to 50 Mbit/s while the remaining two can communicate at up to 25 Mbit/s.

6.2.12 I²S Interface

Five standard I²S interfaces are supported which can operate in master or slave modes, in simplex or full duplex communication modes. Audio sampling frequencies from 8 to 192 kHz are supported.

6.2.13 USART

Three USARTs are supported by the microcontroller. Two of the USARTs can communicate at up to 12.5 Mbit/s, the remaining one can communicate at up to 6.25 Mbit/s.

6.3 SUMMARY

In this chapter, we have learned about the following:

- Key features of the STM32F411RET6 microcontroller
- Power modes
- Electrical characteristics
- ADC and DAC
- Interrupt controller
- Timers
- Clock circuit
- Temperature sensor
- I²C, SPI, and I²S interfaces
- GPIO ports

6.4 EXERCISES

1. Describe the various power modes of the STM32F411RET6 microcontroller.
2. What is the maximum current that can be drawn from a GPIO pin?
3. What is the total maximum current that can be drawn from all the GPIO pins?
4. Describe how we can drive loads requiring more than 20 mA.
5. Explain the various clock sources available on the STM32F411RET6 microcontroller.
6. Explain the difference between push-pull and open-drain output.
7. Explain the difference between pull-up and pull-down pin.
8. What do you understand if a GPIO pin is +5 V tolerant?

7

Using the Mbed With Simple Projects

7.1 OVERVIEW

In this chapter we shall be seeing how to use the Mbed integrated development environment (IDE) to write programs, to compile them, and then how to upload them to the program memory of the Nucleo-F411RE development board.

Additionally, we shall be developing simple projects using Mbed, our development board, and some external components. Each project is described in detail under the following subheadings:

- Project title
- Project description
- Aim of the project
- Project block diagram
- Project circuit diagram (optional)
- Project construction (optional)
- Project algorithm in PDL (optional)
- Project program listing with its description
- Suggestions for additional work (optional)

The algorithm (operational steps) of each program in every project is described using the program description language (PDL), explained in detail in Chapter 2.

7.2 REGISTERING TO USE Mbed

Mbed is a free online ARM compiler that can be used over an Internet link. It is an IDE platform and operating system based on 32-bit ARM Cortex-M microcontrollers. Mbed is supported by over 60 partners and a community of 200,000 developers. Mbed is a free online IDE consisting of an online code editor, a compiler, and a program upload tool. Only a web

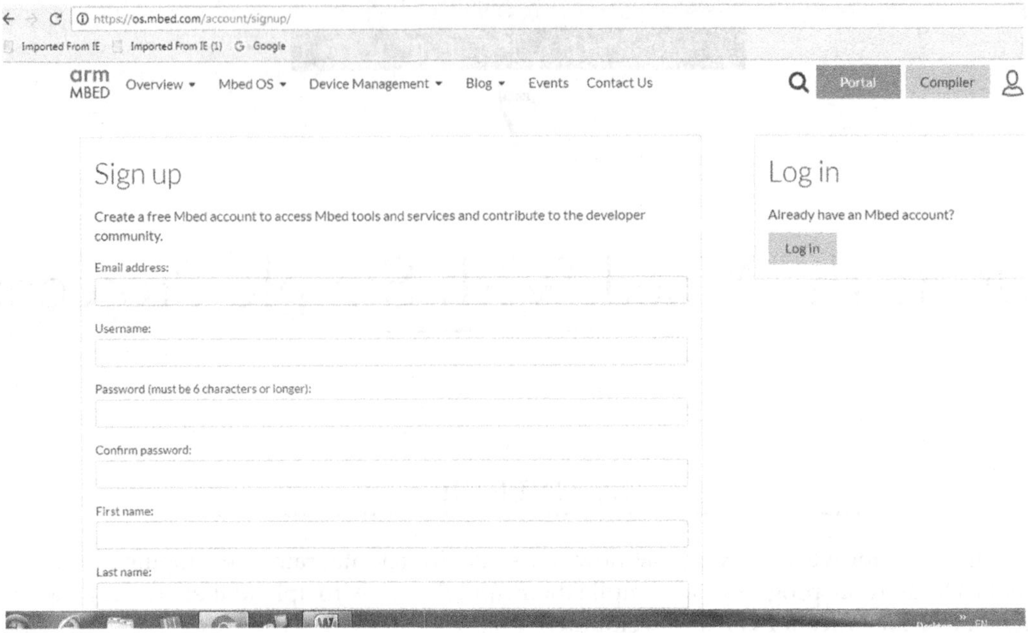

FIG. 7.1 Mbed registration screen.

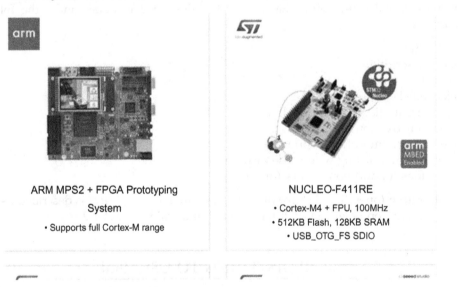

FIG. 7.2 Click on picture Nucleo-F411RE.

browser is required to access Mbed and develop ARM-based programs. Programs are compiled on the Cloud using the ARM C/C++ compiler. Developing a project using Mbed is very easy since all the user needs is to pick an Mbed supported development board, write the application program, and then upload the program to the board.

The Mbed compiler is easy to use and it supports a large number of ARM processors and software libraries. Using Mbed, the compiled code can easily be uploaded to the program memory of the target ARM microcontroller as a simple drag-and-drop (or copy) operation.

First time Mbed users must register before they can use the Mbed and the target ARM-based development board must be added to the Mbed working environment. The steps to register and add the Nucleo-F411RE development board to the Mbed working environment are given in the following:

- Open the following link to create an Mbed login account
   https://os.mbed.com/account/login
- Click **Signup** at the right-hand side and enter your details (Fig. 7.1)
- Click **Hardware -> Boards** and then click on picture Nucleo-F411RE (see Fig. 7.2)
- Click on **Add to your Mbed Compiler** at the right-hand side to add the board to your working environment (Fig. 7.3)
- Click **Compiler** to start the compiler (Fig. 7.4)

① Table of Contents

1. Overview
2. Microcontroller features
3. Board features
4. Board pinout
5. Supported shields
6. Getting started
7. Technical references
8. Known limitations
9. Tips and Tricks

dable and flexible way for users to try out new
M32 microcontroller, choosing from the various
mption and features.

t and the ST Morpho headers allow to expand
o open development platform with a wide choice

ST

A world leader in providing the semiconductor solutions that make a positive contribution to people's lives, both today and in the future.

 Add to your Mbed Compiler

FIG. 7.3   Add Nucleo-F411RE to your working environment.

FIG. 7.4   Start the Mbed compiler.

7.3 NUCLEO-F411RE DEVELOPMENT BOARD PIN NAMES

When working with Mbed we need to know the names of the GPIO pins. These names can be obtained by clicking at the top right-hand side of the Mbed environment where it says NUCLEO-F411RE, or alternatively we can get the pin names from the following site:

https://os.mbed.com/platforms/NUCLEO-F411RE/

The Arduino connector top left and top right pin names are shown in Figs. 7.5 and 7.6, respectively. Similarly, the ST morpho connector top left and top right pin names are shown in Figs. 7.7 and 7.8, respectively.

NUCLEO-F411RE
ARDUINO HEADER
(top left side)

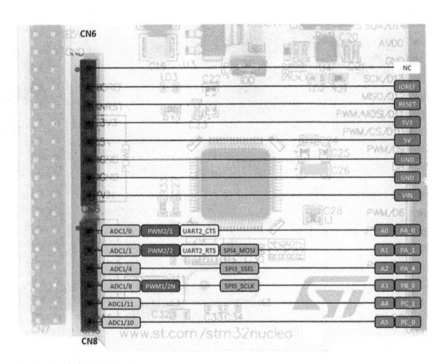

FIG. 7.5   Arduino connector top left pin names.

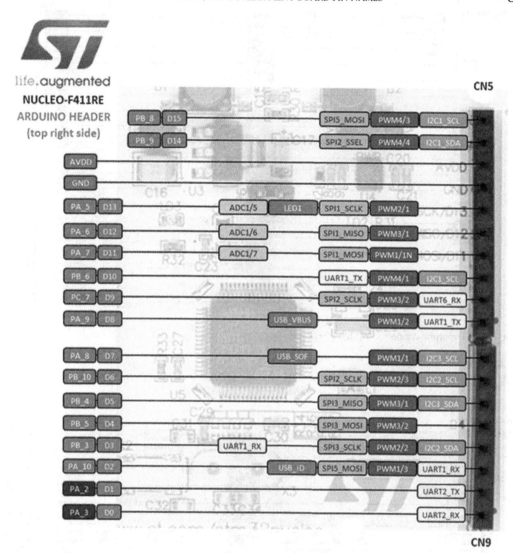

FIG. 7.6   Arduino connector top right pin names.

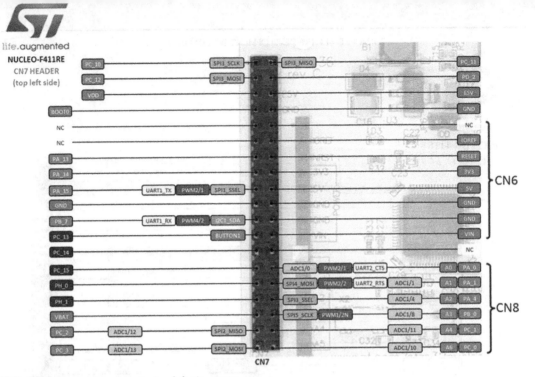

FIG. 7.7   ST morpho connector top left pin names.

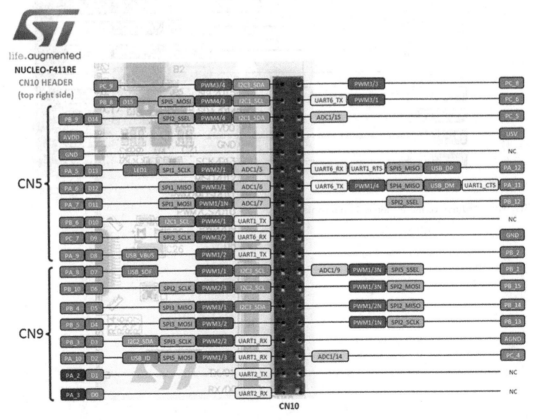

FIG. 7.8   ST morpho connector top right pin names.

7.4 PROJECT 1—FLASHING LED

7.4.1 Description

In this project the user LED on the Nucleo-F411RE board is flashed every 2 s.

7.4.2 Aim

The aim of this project is to show how the Mbed can be used to write a program, compile it, and then upload the program to the program memory of the Nucleo development board. Additionally, the program shows how a GPIO port can be used in output mode and also how the user LED on the board can be flashed at regular intervals.

7.4.3 Block Diagram

Fig. 7.9 shows the block diagram of the project.

7.4.4 PDL of the Project

The PDL of the project is shown in Fig. 7.10

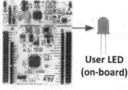

**User LED
(on-board)**

Nucleo-F411RE

FIG. 7.9   Block diagram of the project.

BEGIN
    Configure user LED as output
    DO FOREVER
        Turn ON LED
        Wait 2 seconds
        Turn OFF LED
        Wait 2 seconds
    ENDDO
END

FIG. 7.10   PDL of the project.

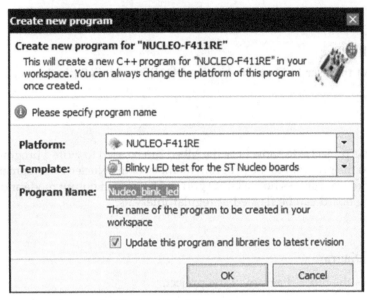

FIG. 7.11   Create a new program called Nucleo_blink_led.

```
/*****************************************************************

            LED FLASHING
            ============

In this program the user LED on the Nucleo-F411RE board (LED1)
is flashed every 2 seconds

Author: Dogan Ibrahim
Date  : August 2018
File  : Nucleo_blink_led
*****************************************************************/
#include "mbed.h"

DigitalOut myled(LED1);

int main()
{
    while(1)
    {
        myled = 1;      // LED is ON
        wait(2.0);      // 2 seconds delay
        myled = 0;      // LED is OFF
        wait(2.0);      // 2 seconds delay
    }
}
```

FIG. 7.12   Program listing.

7.4.5 Program Listing

The steps to create and compile the program are given in the following:

- Login to Mbed site
- Click **Compiler**
- Click **New -> New Program**. Make sure that the Platform is set to **NUCLEO-F411RE**. Click at Template and select **Blinky LED test for the ST Nucleo boards**. Give a name to your program, for example, **Nucleo_blink_led** as shown in Fig. 7.11.
- Expand the **Nucleo_blink_led** listing in the **Program Workspace** (at the left-hand side of the screen). Click on **main.cpp** and then modify the template as shown in Fig. 7.12.
- Click **Compile** to compile the program. After a successful compilation you should see the message **Success!** As shown in Fig. 7.13
- You should also see a pop-up window asking you to Open or Save the compiled code (see Fig. 7.14). Click **Save File** to save the compiled binary file in the download area.
- We are now ready to upload the compiled binary file to the program memory for our Nucleo board. Connect the Nucleo-F411RE board to the USB port of your PC. You should see a new device called NODE_F411RE on your PC (see Fig. 7.15)

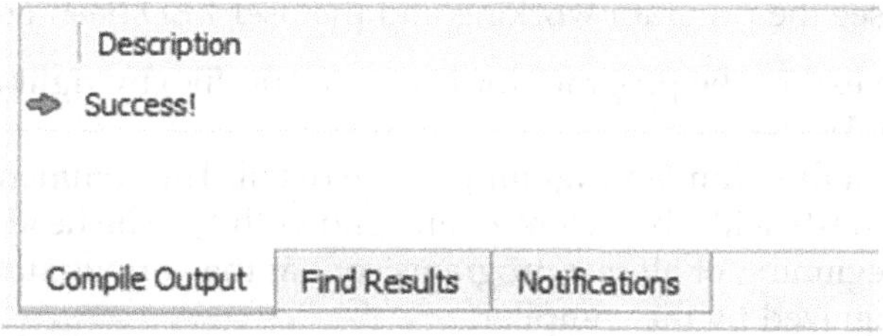

FIG. 7.13   Successful compilation.

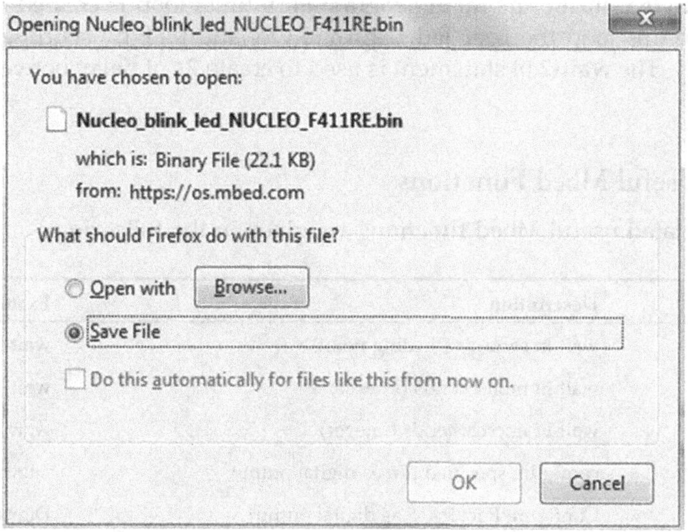

FIG. 7.14   Save the compiled binary file.

FIG. 7.15 New device NODE_F411RE.

- Drag the compiled binary file (the file is called: **Nucleo-blink_led_NUCLEO_F411RE.bin** in this example) from the download area and drop it on device NODE_F411RE. You should see the communications LED on the Nucleo board flashing red and green as the upload process continues. Wait until uploading is complete when the communications LED is steady green color.
- You should now see the program working and the user LED flashing at 2s intervals

Note that you can export the program (or the main.cpp file) by right clicking on the program in the Program Workspace.

We can now look at this simple program in some detail. The comment lines at the beginning of the program starts with characters /* and end with */. The header file **mbed.h** must be included at the beginning of all our programs so that the various statements used in the program can be recognized by the compiler.

The statement **DigitalOut name(pin)** configures the specified pin with the specified name as an output. For example, **DigitalOut myled(LED1)** specifies LED1 to be a digital output with the name **myled**. Inside the main program an infinite loop is created using the **while** statement. Inside this loop the user led is turned ON and OFF by sending 1 and 0 to the LED, respectively. The **wait(2.0)** statement is used to create 2s of delay between each output.

7.4.6 Other Useful Mbed Functions

Some other related useful Mbed functions are given in the following:

Function	Description	Example
wait	wait in seconds (floating point)	wait (2.5)
wait_ms	wait in milliseconds (integer)	wait_ms (10)
wait_us	wait in microseconds (integer)	wait_us (100)
DigitalOut	configure specified pin as digital output	DigitalOut myled(LED1)
DigitalOut	Configure Port PA_5 as digital output	DigitalOut myled(PA_5)

7.4.7 Suggestions for Additional Work

Modify the program in Fig. 7.12 so that the LED ON time is 2 s, and the OFF time is 0.5 s.
The User LED is connected to bit 5 of Port A (PA_5). Modify the program in Fig. 7.12 by
replacing LED1 with PA_5.

7.5 PROJECT 2—LED FLASHING AS MORSE CODE SOS

7.5.1 Description

In this project the user LED on the Nucleo-F411RE board is flashed as Morse code SOS. The
SOS Morse code is "...—...". In this project, a dot is represented with the LED being ON for
0.25 s (Dot time) and a dash is represented with the LED being ON for 1 s (Dash time). The
delay between the dots and dashes is set to 0.2 s (GAP time). This process is repeated continuously after 2 s of delay.

7.5.2 Aim

The aim of this project is to show how for loops can be used in a program. Additionally, the
program shows how to use the **#define** statements in Mbed programs.

7.5.3 Block Diagram

The block diagram of the project is shown in Fig. 7.9.

7.5.4 PDL of the Project

The PDL of the project is shown in Fig. 7.16.

7.5.5 Program Listing

The program listing is shown in Fig. 7.17 (program: **sos**). At the beginning of the program
comment lines are used to describe the operation of the program. Then **mbed.h** is included
and user LED is configured as output and is assigned to name LED. The Dot, Dash, and Gap
times and also ON and OFF are defined at the beginning of the program. The remainder of the
program is executed in an endless loop created using a **while** statement. Inside this loop, two

```
            BEGIN
                    Configure the User LED as output
                    DO FOREVER
                            Flash the LED 3 times with Dot time
                            Wait 0.5 second
                            Flash the LED 3 times with Dash time
                            Wait 2 seconds
                    ENDDO
            END
```

FIG. 7.16   PDL of the project.

```
/******************************************************************
                    LED FLASHING SOS MORSE CODE
                    ===========================

In this program the user LED flashes as an SOS Morse code.
The SOS Morse code is "...---...". Here, a dot is represented
with the LED being ON for 0.25 seconds (Dot time)and a dash is
represented with the LED being ON for 1 second (Dash time. The
delay between the dots and dashes is set to 0.2 second (GAP time).
The process is repeated continuously after 2 seconds of delay

Author: Dogan Ibrahim
Date  : August 2018
File  : sos
******************************************************************/
#include "mbed.h"

DigitalOut LED(LED1);
#define Dot 0.25                      // Dot time
#define Dash 1.0                      // Dash time
#define Gap 0.2                       // Gap time
#define ON 1                          // ON=1
#define OFF 0                         // OFF=0

int main()
{
    int i;
    while(1)                          // Do Forever
    {
        for(i = 0; i < 3; i++)        // Send 3 dots
        {
            LED = ON;                 // LED ON
            wait(Dot);                // Wait Dot time
            LED = OFF;                // LED OFF
            wait(Gap);                // Wait Gap time
        }
        wait(0.5);                    // 0.5 second delay

        for(i = 0; i < 3; i++)        // Send 3 dashes
        {
            LED = ON;                 // LED ON
            wait(Dash);               // Wait Dash time
            LED = OFF;                // LED OFF
            wait(Gap);                // Wait GAp time
        }

        wait(2.0);                    // Wait 2 second before repeating
    }
}
```

FIG. 7.17   Program listing.

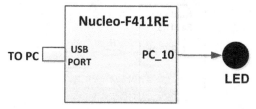

FIG. 7.18   Block diagram of the project.

for loops are used, each looping 3 times. The first **for** loop flashes the LED 3 times where the ON time is set to Dot. The second **for** loop flashes the LED 3 times where the ON time is set to Dash. The end result is that the LED flashes quickly 3 times and then flashes slowly 3 times. This process is repeated forever after 2 s of delay.

7.5.6 Suggestions for Additional Work

Modify the program in Fig. 7.17 so that the flashing stops after sending 10 SOS messages.

7.6 PROJECT 3—FLASHING AN EXTERNAL LED

7.6.1 Description

In this project an external LED is connected to pin 10 of Port C (PC_10) through a current limiting resistor. The project flashes the LED such that the ON time is 0.1 s and the OFF time is 1 s.

7.6.2 Aim

The aim of this project is to show how an external LED can be connected to the Nucleo-F411RE board.

7.6.3 Block Diagram

The block diagram of the project is shown in Fig. 7.18.

7.6.4 Circuit Diagram

An external LED can be connected in two ways: in current source mode and in current sink mode. In both modes the current drawn or supplied by any Nucleo GPIO pin must not exceed 20 mA.

In current source mode the GPIO pin is connected to the LED through a current limiting resistor. The LED turns ON when logic 1 is applied to the corresponding GPIO pin. As shown in Fig. 7.19, the GPIO pin drives the anode pin of the LED through a current limiting resistor

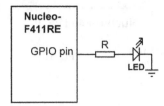

FIG. 7.19   Connecting the LED in current source mode.

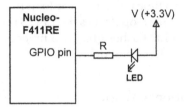

FIG. 7.20   Connecting the LED in current sink mode.

and the cathode pin is connected to the ground. Assuming that the voltage drop across the LED is about 1.8 V and the forward current through the LED is 4 mA, we can calculate the value of the required current limiting resistor as follows:

$$R = V/I$$

Therefore,

$$R = (3.3 - 1.8\,\text{V})/4\text{mA} = 0.375\text{K}$$

where 3.3 V is the output voltage of a GPIO pin when the pin is at logic 1. The nearest physical resistor that we can use 390 ohm (we can use a lower resistor value for higher brightness) which gives a current of

$$I = (3.3 - 1.8\,\text{V})/390 = 3.8\text{mA}$$

In current sink mode the GPIO pin is connected to the LED as shown in Fig. 7.20, where the anode pin of the LED is connected to the supply voltage (+3.3 V) and its cathode is connected to the GPIO pin through a current limiting resistor. The LED turns ON when logic 0 is applied to the corresponding GPIO pin. The value of the required resistor can be calculated as follows:

$$R = (3.3 - 1 - 1.8\,\text{V})/4\text{mA} = 0.125\text{K}$$

where 1 V is the maximum output voltage when the GPIO pin is at logic 0. The nearest physical resistor that we can use 120 ohm (we can use a lower resistor value for higher brightness) which gives a current of

$$I = (3.3 - 1 - 1.8\,\text{V})/120 = 4.1\text{mA}$$

In this project the LED is connected in current source mode with a 390 ohm resistor as shown in Fig. 7.21. GPIO pin PC_10 is at ST morpho connector CN7, and the GND pin is at pin 8 of the same connector (see Fig. 5.8).

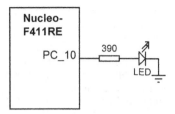

FIG. 7.21   Circuit diagram of the project.

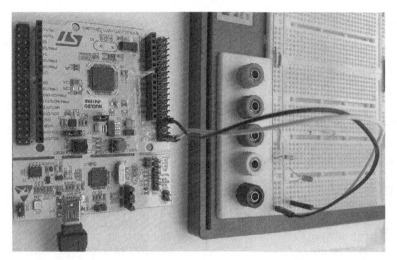

FIG. 7.22   Project constructed on a breadboard.

7.6.5 Project Construction

The project constructed on a breadboard is shown in Fig. 7.22. The LED is connected to the Nucleo board using jumper wires.

7.6.6 Program Listing

Fig. 7.23 shows the program listing (program: **ExtLED**). At the beginning of the program header file **mbed.h** is included. Port PC_10 is configured as output and name LED is assigned to this port pin. Then, the ON and OFF times of the LED are defined, and also ON and OFF are defined. The remainder of the program runs in an endless loop. Inside this loop the LED is flashed with an ON time of 0.1 s and an OFF time of 1 s.

7.6.7 Suggestions for Additional Work

Try using different resistor values and see how the brightness of the LED varies with the resistor values.

```
/************************************************************************
                    EXTERNAL FLASHING LED
                    ====================

In this project an external LED is connected to GPIO port pin 10
of PORT C (PC_10) through a 390 ohm current limiting resistor. The
program flashes the LED quickly such that the ON time is 0.1 second
and the OFF time is 1 second. The net result is that the LED flashes
quickly

Author: Dogan Ibrahim
Date   : August 2018
File   : ExtLED
************************************************************************/
#include "mbed.h"

DigitalOut LED(PC_10);                          // External LED at PC_10

#define ON 1                                    // ON = 1
#define OFF 0                                    // OFF = 0
#define ONTime 0.1                              // ONTime = 0.1s
#define OFFTime 1.0                             // OFFTime = 1.0s

int main()
{
    while(1)                                    // DO FOREVER
    {
        LED = ON;                               // LED ON
        wait(ONTime);                           // Wait ONTime seconds
        LED = OFF;                              // LED OFF
        wait(OFFTime);                          // Wait OFFTime seconds
    }
}
```

FIG. 7.23   Program listing.

7.7 PROJECT 4—ROTATING LEDs

7.7.1 Description

In this project 8 external LEDs are connected to PORT C the Nucleo-F411RE development board. The LEDs turn ON/OFF in a rotating manner every second where only one LED is ON at any time. Fig. 7.24 shows the LED pattern.

7.7.2 Aim

The aim of this project is to show how external LEDs can be connected to the Nucleo-F411RE board.

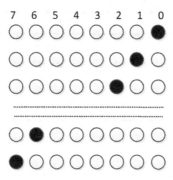

FIG. 7.24   The LED pattern.

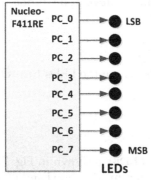

FIG. 7.25   Block diagram of the project.

7.7.3 Block Diagram

The block diagram of the project is shown in Fig. 7.25. The 8 LEDs are connected to PORT C pins of the Nucleo-F411RE development board.

7.7.4 Circuit Diagram

Fig. 7.26 shows the circuit diagram of the project. The LEDs are connected to lower byte of PORT A (GPIO pins PC_0 to PC_7). Current limiting 390 ohm resistors are connected in series with each LED in current sourcing mode.

PORT C pins PC_0 to PC_7 are available at the following ST morpho connector pins (see also Fig. 5.8):

GPIO Pin	ST Morpho Connector
PC_0	CN7, pin 38
PC_1	CN7, pin 36
PC_2	CN7, pin 35
PC_3	CN7, pin 37

GPIO Pin	ST Morpho Connector
PC_4	CN10, pin 34
PC_5	CN10, pin 6
PC_6	CN10, pin 4
PC_7	CN10, pin 19

7.7.5 The Construction

The project was built on a breadboard as shown in Fig. 7.27. Jumper wires are used to make connections between the LEDs and the development board

7.7.6 The PDL

Fig. 7.28 shows the project PDL where the LEDs turn ON in sequence starting from the LSB bit.

7.7.7 Program Listing

The program listing (program: **LED8**) is shown in Fig. 7.29. At the beginning of the program **mbed.h** is included in the program. Then the DigitalOut statement is used to configure

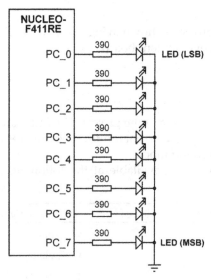

FIG. 7.26　Circuit diagram of the project.

FIG. 7.27   The project constructed on a breadboard.

BEGIN
       Configure used PORT A pins as outputs
       DO FOREVER
            Turn ON LSB LED
            Wait 1 second
            Turn ON next higher LEDs until MSB LED
       ENDDO
    END

FIG. 7.28   PDL of the project.

PORT C pins PC_0 to PC_7 as digital outputs. A **while** statement is used to create an indefinite loop. Inside this loop a **for** statement is used to loop 8 times. The loop counter (variable **k**) is sent to PORT A to turn LEDs ON and OFF as required. One second delay is used between each output.

7.7.8 Modified Program

The program given in Fig. 7.29 can be written using the **PortOut** statement. This statement groups the bits of the given port. The first parameter of this statement is the port name, while the second parameter when set to 1 specifies which bits of the port will be used. In the

```
/************************************************************************

                        ROTATING LEDs
                        =============

In this project 8 LEDs are connected to PORT C pins of the Nucleo-F411RE
development board. The program Turns ON/OFF the LEDs in a rotating manner
such that only one LED is ON at any time 1 second delay is inserted between
each output.

Author: Dogan Ibrahim
Date   : August 2018
File   : LED8
*************************************************************************/
#include "mbed.h"

DigitalOut LEDS[] = {(PC_0),(PC_1),(PC_2),(PC_3),(PC_4),(PC_5),(PC_6),(PC_7)};

int main()
{
    int k;
    while(1)
    {
        for(k = 0; k <= 7; k++)          // Do 8 times
        {
            LEDS[k] = 1;                 // LED ON
            wait(1.0);                   // WAit 1 second
            LEDS[k] = 0;                 // LED OFF
            wait(1.0);                   // WAit 1 second
        }
    }
}
```

FIG. 7.29   Program listing.

modified program listing (program: **LED8-2**) shown in Fig. 7.30, the second parameter of PortOut is set to 0x00FF so that all 8 lower byte of PORT C are used in the program. Array **State** defines the values to be sent to the port so that the required LEDs are turned ON. The program runs indefinitely inside a **while** loop. A **for** statement is used to extract bits from array **State** and send these to the port. In this modified version of the program the delay is set to 250 ms.

7.7.9 Another Modified Program

The program given in this project can be modified by using the BusOut statement as shown in Fig. 7.31 (program: **LED8-3**). The BusOut statement is used to create a group of GPIO pins which can be addressed by a single value. In Fig. 7.31, PORT C pins are grouped as an 8-bit port called LEDS. This port is then accessed by sending a value to it (e.g., LEDS=1 turns ON the first LED in the port group). 0.5 s delay is inserted between each output.

```
**********************************************************************
                    ROTATING LEDS
                    =============

In this modified version of the program The PortOut statement is used
to configure PORT C. Mask 0x00FF sets PC_0 to PC_7 bits to all 1's so
that all the lower byte of PORT C is used.

Author: Dogan Ibrahim
Date   : August 2018
File: LED8-2
**********************************************************************/
#include "mbed.h"

PortOut LEDS(PortC, 0x00FF);
int State[] = {0x01, 0x02, 0x04, 0x08, 0x10, 0x20, 0x40, 0x80};
int main()
{
    while(1)
    {
        for(int k = 0; k < 8; k++)                  // Do 8 times
        {
            LEDS = State[k];                        // Turn LED ON
            wait(0.25);                             // Wait 250ms
        }
    }
}
```

FIG. 7.30   Modified program.

```
/*********************************************************************
                    ROTATE LEDS
                    ===========

In this modified program, the BusOut statement is used. The delay
between teh outputs is set to 0.5 second

Author: Dogan Ibrahim
Date   : August 2018
File   : LED8-3
*********************************************************************/
#include "mbed.h"

BusOut LEDS(PC_0,PC_1,PC_2,PC_3,PC_4,PC_5,PC_6,PC_7);

int main()
{
    int k = 0;                          // First LED
    while(1)                            // Endless loop
    {
        LEDS = 1 << k;                  // LED ON
        wait(0.5);                      // Wait 0.5 second
        k++;                            // Increment LED count
        if(k == 8)k = 0;                // Back to first LED
    }
}
```

FIG. 7.31   Another modified program.

7.8 PROJECT 5—BINARY COUNTING LEDs

7.8.1 Description

In this project 8 external LEDs are connected to PORT C the Nucleo-F411RE development board as in the previous project. The LEDs count up in binary every second as shown in Fig. 7.32.

7.8.2 Aim

The aim of this project is to show how external LEDs can be connected to the Nucleo-F411RE board and how a program can be written so that the LEDs count up in binary.

7.8.3 Block Diagram

The block diagram of the project is shown in Fig. 7.25, where 8 LEDs are connected to PORT C pins of the Nucleo-F411RE development board.

7.8.4 Circuit Diagram

The circuit diagram of the project is shown in Fig. 7.26 where the LEDs are connected using current limiting resistors.

7.8.5 The PDL

Fig. 7.33 shows the PDL of the program.

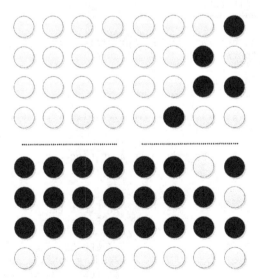

FIG. 7.32   LEDs counting up in binary.

BEGIN
    Group PORT C lower byte
    Set CNT = 1
    DO FOREVER
        Send CNT to PORT C
        Wait 1 second
        Increment CNT
        IF CNT > 255 **THEN**
            CNT = 0
        ENDIF
    ENDDO
END

FIG. 7.33   PDL of the program.

```
/*******************************************************************
                 BINARY UP COUNTER WITH LEDs
                 ===========================

In thsi project 8 LEDs are conencted to PORT C. The LEDs count up
in binary every second.

Author: Dogan Ibrahim
Date   : August 2018
File   : Counter
*******************************************************************/
#include "mbed.h"

BusOut LEDS(PC_0,PC_1,PC_2,PC_3,PC_4,PC_5,PC_6,PC_7);

int main()
{
    int CNT = 0;                        // CNT = 0
    while(1)                            // Endless loop
    {
        LEDS = CNT;                     // Turn ON LED
        wait(1.0);                      // Wait 1 second
        CNT++;                          // Incfement CNT
        if(CNT > 255)CNT = 0;           // CNT back to 0
    }
}
```

FIG. 7.34   Program listing.

7.8.6 Program Listing

The program listing is shown in Fig. 7.34 (program: **Counter**). At the beginning of the program **BusOut** statement is used to group the PORT C lower byte into a variable called LEDS, and variable **CNT** is initialized to 0. The main program is executed in an endless loop using a while statement. Inside this loop the value of **CNT** is sent to PORT C and **CNT** is then incremented by 1. This process is repeated after 1s delay until **CNT** is greater than 255 (i.e., all LEDS are ON) at which point CNT is reset back to 0.

7.9 PROJECT 6—RANDOM FLASHING LEDs

7.9.1 Description

In this project 8 external LEDs are connected to PORT C the Nucleo-F411RE development board as in the previous project. The LEDs turn ON and OFF randomly so that they look like Christmas light. 250ms delay is inserted in every output.

7.9.2 Aim

The aim of this project is to show random numbers can be generated in Mbed programs.

7.9.3 Block Diagram

The block diagram of the project is shown in Fig. 7.25, where 8 LEDs are connected to PORT C pins of the Nucleo-F411RE development board.

7.9.4 Circuit Diagram

The circuit diagram of the project is shown in Fig. 7.26 where the LEDs are connected using current limiting resistors.

7.9.5 The PDL

Fig. 7.35 shows the PDL of the program.

> **BEGIN**
>> Group PORT C lower byte
>> **DO FOREVER**
>>> Generate a random number between 0 and 255
>>> Send the generated number to PORT C
>>> Wait 250ms
>> **ENDDO**
> **END**

FIG. 7.35   PDL of the program.

```
/*********************************************************************
                    RANDOMLY LIGHTING LEDs
                    ======================

in this project 8 LEDs are connecte dto PORT C of the Nucleo-F411RE
development board. Random numbers are generated between 0 and 255 (all
LEDs OFF or all LEDs ON) and the LEDs then turn ON and OFF by sending
these numbers to PORT C every 250ms. The net effect is that the LEDs
look like Christmas light

Author: Dogan Ibrahim
Date  : August 2018
File  : Christmas
*********************************************************************/
#include "mbed.h"

BusOut LEDS(PC_0,PC_1,PC_2,PC_3,PC_4,PC_5,PC_6,PC_7);

int main()
{
    int Number;

    while(1)                                // Do forever
    {
        Number = rand() % 256;              // GEnerate a random numner
        LEDS = Number;                      // Send the numebr to PORT C
        wait(0.25);                         // 250ms delay
    }
}
```

FIG. 7.36   Program listing.

7.9.6 Program Listing

The program listing is shown in Fig. 7.36 (program: **Christmas**). At the beginning of the program **BusOut** statement is used to group the PORT C lower byte into a variable called LEDS as in the previous project. An endless loop is created using a **while** statement. Inside this loop, random numbers are generated between 0 and 255 using function **rand()**, and these numbers are then sent to PORT C every 250 ms The net effect is that the Les turn ON and OFF as Christmas lights.

7.9.7 Modified Program

The program given in Fig. 7.36 can be modified by making the delay between the outputs to change randomly. This gives the effect of randomly flashing lights with random delays. In the modified program in Fig. 7.37 (program: **Christmas-2**) the delay between the outputs changes randomly between 0 and 300 ms.

```
/****************************************************************

                   RANDOMLY FLASHING LEDs
                   =======================

In this version of teh program the delay between the outputs
is random and varies between 0 and 300ms.

Author: Dogan Ibrahim
Date   : August 2018
File   : Christmas-2
****************************************************************/
#include "mbed.h"

BusOut LEDS(PC_0,PC_1,PC_2,PC_3,PC_4,PC_5,PC_6,PC_7);


int main()
{
    int Number, Tim;
    float Del;

    while(1)
    {
        Number = rand() % 256;          // Random number 0-255
        Tim = rand() % 11;              // Random number 0 - 10
        Del = Tim *0.03;                // Random delay 0-300ms
        LEDS = Number;                  // Turn ON LEDs
        wait(Del);                      // Random delay
    }
}
```

FIG. 7.37 Modified program.

7.10 PROJECT 7—LED CONTROL WITH PUSH-BUTTON

7.10.1 Description

   In this project the user push-button and the user LED on the Nucleo-F411RE development board are used. The LED flashes 3 times when the button is pressed.

7.10.2 Aim

   The aim of this project is to show how the user push-button on the Nucleo board can be used.

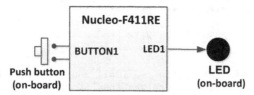

FIG. 7.38   Block diagram of the project.

```
BEGIN
        Configure LED1 as output
        Configure BUTTON1 as input
        DO FOREVER
                IF BUTTON1 is pressed THEN
                        Flash LED1 3 times
                ENDIF
        ENDDO
END
```

FIG. 7.39   PDL of the program.

7.10.3 Block Diagram

The block diagram of the project is shown in Fig. 7.38. The User LED (at PB_13) is named LED1 and the User Button (at PC_13) is named BUTTON1 (see Fig. 5.8). The User Button is normally at logic 1 and it goes to logic 0 when pressed. The User LED turns ON when logic 1 is applied to it.

7.10.4 The PDL

The PDL of the program is very simple and is shown in Fig. 7.39.

7.10.5 Program Listing

The program listing (program: **Push**) is shown in Fig. 7.40. At the beginning of the program the header file **mbed.h** is included in the program. The User LED (LED1) is configured as an output using the **DigitalOut** statement. Similarly, the User Button (BUTTON1) is configured as an input using the **DigitalIn** statement. The remainder of the program runs in an endless loop setup using a **while** statement. The program checks if the button is pressed (i.e., button is at logic 0) and if so a **for** loop is used to flash the LED 3 times.

```
/********************************************************************

                    PUSH BUTTON WITH LED
                    ====================

In this program teh user Button (BUTTON1) and the User LED (LED1)
are used. When the button is pressed the LED flashed 3 times with
one second delay between each output.

Author: Dogan Ibrahim
Date   : August 2018
File   : Push
********************************************************************/
#include "mbed.h"

DigitalOut led(LED1);                       // LED1 isoutput
DigitalIn button(BUTTON1);                  // BUTTON1 is input

int main()
{
    while(1)                                // DO forever
    {
        if(button == 0)                     // If button is pressed
        {
            for(int k = 0; k < 3; k++)      // Do 3 times
            {
                led = 1;                    // LED ON
                wait(1.0);                  // Wait 1 second
                led = 0;                    // LED OFF
                wait(1.0);                  // Wait 1 second
            }
        }
    }
}
```

FIG. 7.40   Program listing.

7.11 PROJECT 8—CHANGING LED FLASHING RATE

7.11.1 Description

In this project the user push-button and the user LED on the Nucleo-F411RE development board are used. The LED flashing rate is changed every time the button is pressed. This is done by changing the delay between the flashes.

7.11.2 Aim

The aim of this project is to show how the user push-button on the Nucleo board can be used.

BEGIN
      Configure LED1 as output
      Configure BUTTON1 as input
      DO FOREVER
          IF BUTTON1 is pressed **THEN**
             Change delay time
             Flash the LED with current delay time
          ENDIF
      ENDDO
   END

FIG. 7.41   PDL of the program.

7.11.3 Block Diagram

The block diagram of the project is shown in Fig. 7.38.

7.11.4 The PDL

The PDL of the program is very simple and is shown in Fig. 7.41.

7.11.5 Program Listing

The program listing (program: **LEDrate**) is shown in Fig. 7.42. At the beginning of the program the header file **mbed.h** is included in the program. The User LED (LED1) is configured as an output using the **DigitalOut** statement. Similarly, the User Button is configured as an input using the **DigitalIn** statement. The delay between the flashes (variable **dely**) is initially set to 1 s. Every time the button is pressed this delay is decremented by 200 ms and as a result the LED flashes quicker. When the delay is less than 0, it is reset back to 1 s.

7.12 PROJECT 9—BINARY EVENT COUNTING WITH LEDs

7.12.1 Description

In this project 8 external LEDs are connected to PORT C the Nucleo-F411RE development board as in Project 4. The LEDs count up by one in binary every time the User button is pressed.

7.12.2 Aim

The aim of this project is to show how the User button can be used to simulate external events and also how the count of such events can be displayed in binary on 8 LEDs.

```
/*********************************************************************

                    CHANGING LED FLASHING RATE
                    ==========================

In this program the User LED and the Usser button are used. The flashing
rate of teh LED starts from 1 second and is then increased by 200ms every
time the button is pressed. When the flashing rate drops to zero it is
reset back to one second

Author: Dogan Ibrahim
Date  : August 2018
File  : LEDrate
**********************************************************************/
#include "mbed.h"

DigitalOut led(LED1);                        // LED is output
DigitalIn button(BUTTON1);                   // Button is input


int main()
{
    float dely = 1.0;                        // Starting delay

    while(1)                                 // Do forever
    {
        led = 1;                             // LED ON
        wait(dely);                          // Wait dely seconds
        led = 0;                             // LED OFF
        wait(dely);                          // Wait dely seconds
        if(button == 0)                      // If button is pressed
        {
            dely = dely - 0.2;               // decrement dely
            if(dely < 0)dely = 1.0;          // if delay is < 0
        }
    }
}
```

FIG. 7.42   Program listing.

7.12.3 Block Diagram

The block diagram of the project is shown in Fig. 7.43, where 8 LEDs are connected to PORT C pins of the Nucleo-F411RE development board.

7.12.4 Circuit Diagram

The circuit diagram of the project is shown in Fig. 7.44 where the LEDs are connected using current limiting resistors. The on-board User button is used to create external events.

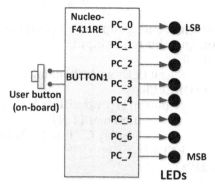

FIG. 7.43 Block diagram of the project.

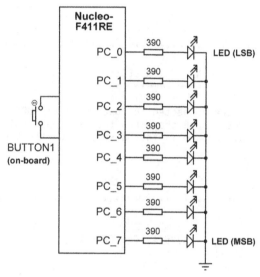

FIG. 7.44 Circuit diagram of the project.

7.12.5 The PDL

Fig. 7.45 shows the PDL of the program.

7.12.6 Program Listing

The program listing is shown in Fig. 7.46 (program: **Events**). At the beginning of the program **BusOut** statement is used to group the PORT C lower byte into a variable called **LEDS** and variable **Cnt** is initialized to 0 and all the LEDs are turned OFF to start with. The remainder of the program runs in an endless loop. Inside this loop the state of the button is checked. If the button is pressed (button is 0) then variable **Cnt** is incremented by one and the total count is displayed in binary on the LEDs. The program then waits until the button is released and the above process is repeated.

BEGIN
    Configure and group LEDs as outputs
    Configure button as input
    Cnt = 0
    Turn OFF all LEDs at the beginning
    DO FOREVER
        IF button is pressed **THEN**
            Increment CNT
            Display CNT on LEDs in binary
        ENDIF
    ENDDO
END

FIG. 7.45   PDL of the program.

```
/***************************************************************************

                BINARY EVENT COUNTER WITH LEDS
                ==============================

In this program 8 LEDs are connected to lower byte of PORT C. The program
simulates an ebent counter such that when the User button is pressed a
count is incremented by one and is then displayed on the LEDs.

Author: Dogan Ibrahim
Date  : August 2018
File  : Events
***************************************************************************/
#include "mbed.h"

BusOut LEDS(PC_0,PC_1,PC_2,PC_3,PC_4,PC_5,PC_6,PC_7);
DigitalIn button(BUTTON1);

int main()
{
    int Cnt = 0;                            // Cnt = 0 to start with
    LEDS = 0;                               // All LEDs OFF at beginning

    while(1)                                // DO Forever
    {
        if(button == 0)                     // If button is pressed
        {
            Cnt++;                          // Increment Cnt
            LEDS = Cnt;                     // Display Cnt
            while(button == 0);             // Wait until button released
        }
    }
}
```

FIG. 7.46   Program listing.

7.13 PROJECT 10—USING AN EXTERNAL BUTTON

7.13.1 Description

In this project an external button is connected to port pin PC_0 (connector CN7, pin 38) of the 8 Nucleo-F411RE development board. The User LED (LED1) is turned ON whenever the button is pressed.

7.13.2 Aim

The aim of this project is to show how an external button can be connected to the Nucleo-F411RE development board.

7.13.3 Block Diagram

The block diagram of the project is shown in Fig. 7.47.

7.13.4 Circuit Diagram

Buttons can be connected in two different ways to a microcontroller input port. In Fig. 7.48 the button state is normally held at logic 1 with the help of an external pull-up resistor and it goes to logic 0 when the button is pressed. In Fig. 7.49 the button state is normally at logic 0 and goes to logic 1 when the button is pressed. It is possible to pull-up a GPIO input pin internally via a pull-up resistor. In such applications there is no need to use an external pull-up resistor.

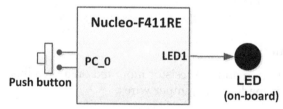

FIG. 7.47 Block diagram of the project.

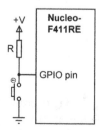

FIG. 7.48 Button state is normally at logic 1.

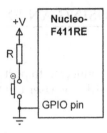

FIG. 7.49   Button state is normally at logic 0.

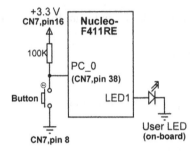

FIG. 7.50   Circuit diagram of the project.

In this project the first method is used, that is, the button state is normally at logic 1 and goes to logic 0 when the button is pressed. The circuit diagram of the project is shown in Fig. 7.50. The button is connected to the following pins of the Nucleo board:

+3.3 V to connector CN7, pin 16
GND to connector CN7, pin 8
PC_0 to connector CN7, pin 38

7.13.5 The Construction

Fig. 7.51 shows the button and the resistor mounted on a breadboard. Connection to the Nucleo-F411RE board is made using jumper wires.

7.13.6 The PDL

Fig. 7.52 shows the PDL of the program.

7.13.7 Program Listing

The program listing is shown in Fig. 7.53 (program: **EXTbutton**). At the beginning of the program **DigitalOut** and **DigitalIn** statements are used to configure the LED and the button as digital output and input, respectively. The remainder of the program runs in an endless loop where the state of the button is checked continuously. Whenever the button is pressed the LED is turned ON, otherwise the LED is turned OFF.

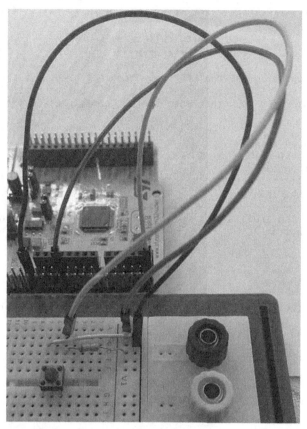

FIG. 7.51   Project built on a breadboard.

BEGIN
   Configure Button as input
   Configure LED as output
   DO FOREVER
       IF button is pressed **THEN**
           Turn ON LED
       ELSE
           Turn OFF LED
       ENDIF
   ENDDO
END

FIG. 7.52   PDL of the program.

7.13.8 DigitalIn Modes

The DigitalIn statement supports the following modes:

PullUp: Pull-up the GPIO pin internally (no need for external resistor)
PullDown: Pull-down the GPIO pin

```
/****************************************************************

                    EXTERNAL BUTTON WITH LED
                    ========================

In this program an external button is conencted to GPIO pin PC_0
of teh Nucleo-F411RE development board. This button controls the
on-board User LED such that when the button is pressed the LED
turns ON

Author: Dogan Ibrahim
Date   : August 2018
File   : EXTbutton
****************************************************************/
#include "mbed.h"

DigitalOut led(LED1);                       // LED is output
DigitalIn button(PC_0);                     // Button is input

int main()
{
    while(1)                                // Do forever
    {
        if(button == 0)                     // If button pressed
            led = 1;                        // LED ON
        else
            led = 0;                        // LED OFF
    }
}
```

FIG. 7.53   Program listing.

PullNone: No pull-up/pull-down resistors
OpenDrain: GPIO pin on open-drain mode

The earlier modes can either be specified during the declaration of the DigitalIn statement,or later as shown in the following:
    DigitalIn button(PC_0, PullUp);
or as:
    DigitalIn button(PC_0);
    button.mode(PullUp);

7.13.9 Suggestions for Additional Work

Modify the circuit in Fig. 7.50 by removing the external pull-up resistor and enable the internal pull-up resistor on port pin PC_0 in the program as shown in Fig. 7.53.

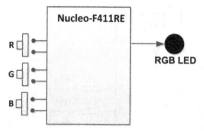

FIG. 7.54   Block diagram of the project.

FIG. 7.55   Common-cathode RGB LED.

7.14 PROJECT 11—RGB LED CONTROL

7.14.1 Description

In this project an RGB LED is connected to the Nucleo-F411RE development board. Additionally, three push-buttons named as R, G, and B are connected to the board corresponding to three colors Red, Green, and Blue. Pressing a button toggles the state of the corresponding color. For example, if the Red color is OFF, then pressing button R will turn ON this color. If on the other hand the Red color is ON, then pressing button R will turn OFF this color.

7.14.2 Aim

The aim of this project is to show how more than one external button can be connected to the Nucleo board. Additionally, the project shows how an RGB LED can be used in a project.

7.14.3 Block Diagram

The block diagram of the project is shown in Fig. 7.54.

7.14.4 Circuit Diagram

The RGB LED (see Fig. 7.55) is a 4-pin LED which can display the mixture of the Red, Green, and Blue colors depending on the pin that has been activated. These LEDs can be either in common-anode or in common-cathode configurations. The one that will be used in this project is a common-cathode RGB LED where the common pin is connected to ground and the individual colors are displayed when the corresponding pins are set to logic 1. As shown in Fig. 7.55, the common pin is longer than the other pins for identification.

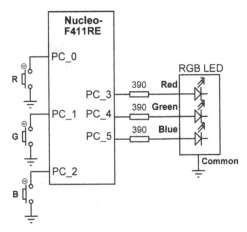

FIG. 7.56   Circuit diagram of the project.

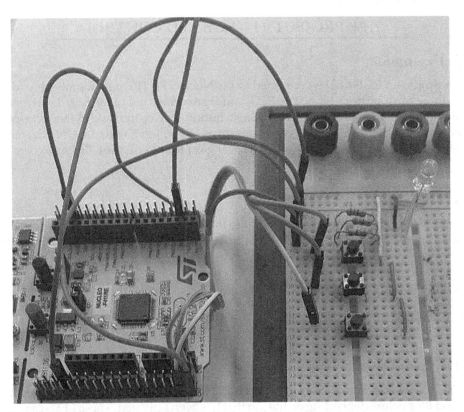

FIG. 7.57   The project constructed on a breadboard.

Fig. 7.56 shows the circuit diagram of the project. The three buttons R, G, and B are connected to GPIO pins PC_0, PC_1, and PC_2, respectively. The Red, Green, and Blue pins of the LED are connected to GPIO pins PC_3, PC_4, and PC_5 through 390 ohm current limiting resistors, respectively.

7.14.5 The Construction

Fig. 7.57 shows the project build on a breadboard. The RGB LED is connected to the Nucleo-F411RE board using jumper wires.

7.14.6 The PDL

Fig. 7.58 shows the program PDL.

7.14.7 Program Listing

The program listing (program: **RGB**) is shown in Fig. 7.59. At the beginning of the program **BusOut** statement is used to group the RGB control pins, and **DigitalIn** statements are used to configure the three buttons R,G, and B as inputs. All these three colors are turned OFF at the beginning of the program. The program then checks if a button is pressed and if so checks to see if the color corresponding to that button is already displayed. If the color is already

```
BEGIN
        Configure the RGB as output
        Configure R,G,B button as inputs and enable Pull-Ups
        Clear all LEDs
        DO FOREVER
                IF R is pressed THEN
                        IF Red colour is displayed THEN
                                Remove Red colour
                        ELSE
                                Add Red colour
                ENDIF
                IF G is pressed THEN
                        IF Green colour is displayed THEN
                                Remove Green colour
                        ELSE
                                Add Green colour
                ENDIF
                IF B is pressed THEN
                        IF Blue colour is displayed THEN
                                Remove Blue colour
                        ELSE
                                Add Blue colour
                        ENDIF
                ENDIF
        ENDDO
END
```

FIG. 7.58   PDL of the project.

```
/*************************************************************************

                          RGB LED CONTROL
                          ===============

In this project an RGB LED is connected to the Nucleo board. Additionally,
3 buttons named R,G, and B are conencted to the board. When a button is
pressed the corresponding colour is displayed by the RGB LED. Pressing the
same button again removes that colour from the display.

The buttons are pulled-up in software and are connected as follows:
R to PC_0
G to PC_1
B to PC_2

The RGB LED pins are connected as follows:
Red   to PC_3
Green to PC_4
Blue  to PC_5

Author: Dogan Ibrahim
Date  : August 2018
File  : RGB
*************************************************************************/

#include "mbed.h"

BusOut RGB(PC_3, PC_4, PC_5);                       // Group the LED pins
DigitalIn R(PC_0, PullUp);                          // R button
DigitalIn G(PC_1, PullUp);                          // G button
DigitalIn B(PC_2, PullUp);                          // B button

int main()
{
    RGB = 0;                                        // LEDs OFF at beginning

    while(1)                                        // Do forever
    {
        if(R == 0)                                  // If R pressed
        {
            if(RGB & 1 == 1)                        // If Red is active
                RGB = RGB & 6;                      // Turn OFF Red
            else                                    // Otherwise
                RGB = RGB | 1;                      // Turn ON Red
            while(R == 0);                          // Wait until released
        }
        else if(G == 0)                             // If G pressed
        {
            if((RGB & 2) == 2)                      //If Green is active
                RGB = RGB & 5;                      // Turn OFF Green
            else                                    // Otherwise
                RGB = RGB | 2;                      // Turn ON Green
            while(G == 0);                          // Wait until released
        }
```

FIG. 7.59 Program listing.

(Continued)

```
        else if(B == 0)                        // If B is pressed
        {
            if((RGB & 4) == 4)                 // If Blue is active
                RGB = RGB & 3;                 // Turn OFF Blue
            else                               // Otherwise
                RGB = RGB | 4;                 // Turn ON Blue
            while(B == 0);                     // Wait until released
        }
    }
}
```

FIG. 7.59, CONT'D

displayed then it is removed, otherwise it is displayed. This is repeated for all the three buttons.

The Red, Green, and Blue LEDs are connected to PORT C in the following order:

Blue Green Red

The relationship between the RGB value and the colors that turn ON are as follows:

RGB	Action
1	Turn ON Red
2	Turn ON Green
4	Turn ON Blue
6	Turn OFF Red
5	Turn OFF Green
3	Turn OFF Blue

7.15 PROJECT 12—RANDOM COLORED LED DISPLAY

7.15.1 Description

In this project an RGB LED is connected to the Nucleo-F411RE development board. The three colors Red, Green, and Blue of the RGB are activated randomly every 250 ms. The net effect is that the LED displays different colors randomly.

7.15.2 Aim

The aim of this project is to show how an RGB LED can be activated randomly.

7.15.3 Block Diagram

The block diagram of the project is shown in Fig. 7.60.

7.15.4 Circuit Diagram

Fig. 7.61 shows the circuit diagram of the project. The Red, Green, and Blue pins of the RGB LED are connected to GPIO pins PC_3, PC_4, and PC_5, respectively.

7.15.5 The Construction

The project is constructed on a breadboard as shown in Fig. 7.62.

7.15.6 The PDL

Fig. 7.63 shows the PDL of the program.

7.15.7 Program Listing

The program listing (program: **RGB-2**) is shown in Fig. 7.64. At the beginning of the program the **BusOut** statement is used to group the LED pins into a variable called RGB. Then, an

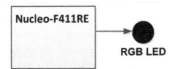

FIG. 7.60   Block diagram of the project.

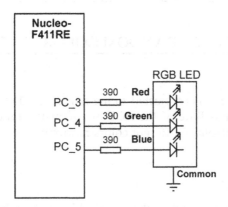

FIG. 7.61   Circuit diagram of the project.

FIG. 7.62 Project constructed on a breadboard.

BEGIN
        Group RGB LED pins as a variable
        DO FOREVER
                Create a random number between 1 and 4
                Send the generated number to RGB LED
                Wait 250ms
        ENDDO
END

FIG. 7.63 PDL of the program.

endless loop is formed where a random number is generated between 1 and 4 and this number is sent to the LED to display different colors every 250ms.

7.15.8 Suggestions for Additional Work

Modify the program given in Fig. 7.64 by randomly varying the delay between 50 and 300ms. What are the effects of modifying this delay?

```
/*****************************************************************

               RANDOM COLOURED LED DISPLAY
               ===========================

In this project an RGB LED is connected to the Nucleo-F411RE
development board. The Red, Green, and Blue pins of the LED are
connected to GPIO pins PC_3, PC_4 and PC_5 respectively. The LED
displays random colours every 250ms.

Author: Dogan Ibrahim
Date   : August 2018
File   : RGB-2
*****************************************************************/
#include "mbed.h"

BusOut RGB(PC_3,PC_4,PC_5);                 // Group the LED pins

int main()
{
    int Number;

    while(1)                                // Do forever
    {
        Number = rand() % 5;                // GEnerate a random number
        RGB = Number;                       // Send the number to the LED
        wait(0.25);                         // Wait 250ms
    }
}
```

FIG. 7.64   Program listing.

7.16 PROJECT 13—7-SEGMENT LED DISPLAY

7.16.1 Description

In this project a single-digit 7-segment display is interfaced to the Nucleo-F411RE development board. The display counts up every second from 0 to 9.

7.16.2 Aim

The aim of this project is to show how a 7-segment display can be interfaced to the Nucleo-F411RE development board. Additionally, the project shows how to control such a display by programming using Mbed.

7.16.3 Block Diagram

The block diagram of the project is shown in Fig. 7.65.

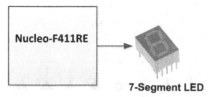

FIG. 7.65 Block diagram of the project.

FIG. 7.66 Some 7-Segment LEDs.

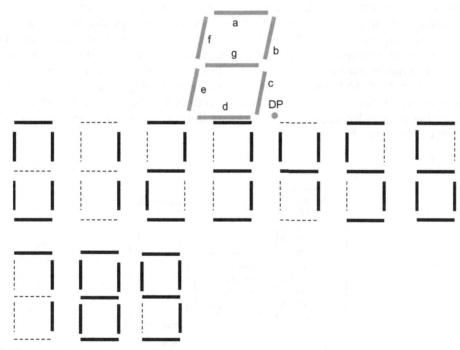

FIG. 7.67 Segments of a 7-Segment LED.

7-segment displays (see Fig. 7.66) are used in microcontroller-based applications to display numbers and some letters. A 7-segment display consists of 7 LEDs connected such that the numbers 0–9 and some letters can be displayed by turning the appropriate LED segments ON or OFF.

The segments of a 7-segment LED are identified by letters **a–g** (or **A–G**) s shown in Fig. 7.67.

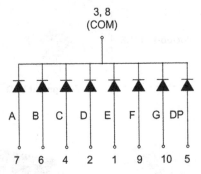

FIG. 7.68   SMA42056 7-Segment display pins.

7-Segment displays are available as either common anode or common cathode configurations. In common anode displays all the anodes of the segments are connected together and this common pin is connected to a power supply. A segment is turned ON when its corresponding cathode pin is at logic 0. In a common cathode type display all the cathode pins of the segments are connected together and this common pin is usually connected to ground. A segment is turned ON when its corresponding anode pin is at logic 1. In addition to displaying numeric values, a 7-segment LED can also display a decimal point.

In this project the SMA42056 type common cathode display is used. This is a 0.56 in. (14.20 mm) red color display. Fig. 7.68 shows the connection diagram of this display. Pin 1 starts from the bottom left corner of the display. Bottom right-hand corner is pin 5, and the top left corner is pin 10.

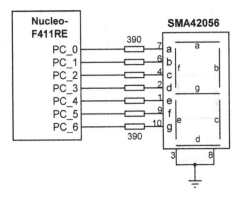

FIG. 7.69   Circuit diagram of the project.

7.16.4 Circuit Diagram

The circuit diagram of the project is shown in Fig. 7.69. PORT C lower byte pins (PC_0 to PC_6) are connected to the 7-segment pins via 390 ohm current limiting resistors as follows:

Segment	GPIO Pin	ST Morpho Connector
a	PC_0	CN7, pin 38
b	PC_1	CN7, pin 36
c	PC_2	CN7, pin 35
d	PC_3	CN7, pin 37
e	PC_4	CN10, pin 34
f	PC_5	CN10, pin 6
g	PC_6	CN10, pin 4

The relationship between the numbers to be displayed and the segments that should be turned ON are given in the following:

Number to be Displayed	Segments to be ON
0	a b c d e f
1	b c
2	a b g e d
3	a b g c d
4	f g c b
5	a f g c d
6	a f g e d c
7	a b c
8	a b c d e f g
9	a b g f c d

Before writing the program we need to know the relationship between the numbers to be displayed and the data that should be sent to PORT C to display the requested numbers. This is shown in Table 7.1. Note that PC_7 is not used in the 7-segment display and it is set to 0 in this table for convenience.

TABLE 7.1 Number to be Displayed ad PORT C Value

Number	PC_7	PC_6	PC_5	PC_4	PC_3	PC_2	PC_1	PC_0	Hex
0	0	0	1	1	1	1	1	1	0x3F
1	0	0	0	0	0	1	1	0	0x06
2	0	1	0	1	1	0	1	1	0x5B
3	0	1	0	0	1	1	1	1	0x4F
4	0	1	1	0	0	1	1	0	0x66
5	0	1	1	0	1	1	0	1	0x6D
6	0	1	1	1	1	1	0	1	0x7D
7	0	0	0	0	0	1	1	1	0x07
8	0	1	1	1	1	1	1	1	0x7F
9	0	1	1	0	1	1	1	1	0x6F

Number	PORT C Value
0	0x3F
1	0x06
2	0x5B
3	0x4F
4	0x66
5	0x6D
6	0x7D
7	0x07
8	0x7F
9	0x6F

As we can see from Table 7.1, the following hexadecimal number must be sent to the lower byte of PORT C in order to display a number:

7.16.5 The Construction

Fig. 7.70 shows the project built on a breadboard.

7.16.6 The PDL

Fig. 7.71 shows the program PDL.

FIG. 7.70   Project built on a breadboard.

BEGIN
>Configure PORT C lower byte as a group
>Define segment patterns
>CNT = 0
>**DO FOREVER**
>>Send CNT to display
>>Wait 1 second
>>**IF** CNT = 10 **THEN**
>>>CNT = 0
>>**ENDIF**
>**ENDDO**

END

FIG. 7.71   Program PDL.

7.16.7 Program Listing

The program listing (program: **SevenSeg**) is shown in Fig. 7.72. At the beginning of the program PORT C lower byte is grouped into variable **Segments** using the **PortOut** statement. The segments to be turned ON for a required number to be displayed are given in array **LEDS**.

```
/*************************************************************************

                            7-SEGMENT COUNTER
                            =================

    In thsi project a 7-Segment LED is conencted to lower byte of PORT C
    ofthe Nucleo-F411RE development board. The program counts from 0 to 9
    every second

    Author: Dogan Ibrahim
    Date   : August 2018
    File   : SevenSeg
    **********************************************************************/
#include "mbed.h"

PortOut Segments(PortC, 0xFF);
int LEDS[] = {0x3F,0x06,0x5B,0x4F,0x66,0x6D,0x7D,0x07,0x7F,0x6F};


int main()
{
    int CNT = 0;                            // Initialize CNT

    while(1)                                // Do forever
    {
        Segments = LEDS[CNT];               // Send CNT to LED
        wait(1.0);                          // Wait 1 second
        CNT++;                              // Increment CNT
        if(CNT == 10)CNT = 0;               // Reset CNT to 0
    }
}
```

FIG. 7.72   Program listing.

For example, to display number 0 we have to send the hexadecimal number 0x3F to the LED. Similarly, to display number 1 we have to send the hexadecimal number 0x06 to the LED and so on. Inside the program loop array **LEDS** is indexed by variable **CNT** and the resulting segment pattern is sent to PORT C. Variable **CNT** is then incremented by 1 and the program waits for 1 s. When **CNT** reaches to 10 it is reset back to 0. The above process repeats forever displaying numbers 0–9 on the 7-segment LED.

7.16.8 Modified Program

The program given in Fig. 7.72 can be written differently using the **BusOut** statement. This modified program (program: **SevenSeg-2**) is shown in Fig. 7.73. Used PORT C pin names are defined in statement **BusOut**. The remainder of the program is same as in Fig. 7.72.

```
/*********************************************************************

          7-SEGMENT LED COUNTER
          =====================

In this modified versiuon of the program the BusOut statement is used

Author: Dogan Ibrahim
Date   : August 2018
File   : SevenSeg-2
*********************************************************************/
#include "mbed.h"

BusOut Segments(PC_0,PC_1,PC_2,PC_3,PC_4,PC_5,PC_6);
int LEDS[] = {0x3F,0x06,0x5B,0x4F,0x66,0x6D,0x7D,0x07,0x7F,0x6F};

int main()
{
    int CNT = 0;                        // Initialize CNT

    while(1)                            // DO forever
    {
        Segments = LEDS[CNT];           // Send CNT to LED
        wait(1.0);                      // Wait 1 second
        CNT++;                          // Increment CNT
        if(CNT == 10)CNT = 0;           // Reset CNT to 0
    }
}
```

FIG. 7.73   Modified program.

7.16.9 Another Modified Program

The program given in Fig. 7.72 can be modified using a function as shown in Fig. 7.74. In the modified program (program: **SevenSeg-3**) function **Display** receives the number to be displayed as its argument and then sends this number to the LED to display it.

7.16.10 Using Switch Statement

The program given in Fig. 7.74 can be modified using a **switch** statement. The modified program (program: **SevenSeg-4**) is shown in Fig. 7.75. Here, a **switch** statement is used inside function **Display** to display the required number on the 7-segment LED.

```
/**********************************************************************

                    7-SEGMENT LED COUNTER
                    =====================

In this evrsion of teh program a function is used to display the number

Author: Dogan Ibrahim
Date   : August 2018
File   : SevenSeg-3
**********************************************************************/
#include "mbed.h"

PortOut Segments(PortC, 0xFF);
int LEDS[] = {0x3F,0x06,0x5B,0x4F,0x66,0x6D,0x7D,0x07,0x7F,0x6F};

//
// This function displays number N on the 7-Segment LED
//
void Display(int N)
{
    Segments = LEDS[N];
}


//
// MAIN program
//
int main()
{
    int CNT = 0;

    while(1)
    {
        Display(CNT);
        wait(1.0);
        CNT++;
        if(CNT == 10)CNT = 0;
    }
}
```

FIG. 7.74   Another modified program.

7.16.11 Suggestions for Additional Work

Modify the program given in Fig. 7.72 to display letters L and E alternately on the 7-segment display.

```
/*****************************************************************
                    7-SEGMENT LED COUNTER
                    =====================

In this modified program a switch statement is used inside a
function to display the required number

Author: Dogan Ibrahim
Date  : Augut 2018
File  : SevenSeg-4
*****************************************************************/
#include "mbed.h"

PortOut Segments(PortC, 0xFF);

//
// This function displays the requirednumber
//
void Display(int N)
{
     switch (N)
        {
            case 0:
                Segments = 0x3F;
                break;
            case 1:
                Segments = 0x06;
                break;
            case 2:
                Segments = 0x5B;
                break;
            case 3:
                Segments = 0x4F;
                break;
            case 4:
                Segments = 0x66;
                break;
            case 5:
                Segments = 0x6D;
                break;
            case 6:
                Segments = 0x7D;
                break;
            case 7:
                Segments = 0x07;
                break;
```

FIG. 7.75 Program using switch statement.

(Continued)

```
                    case 8:
                        Segments = 0x7F;
                        break;
                    case 9:
                        Segments = 0x6F;
                        break;
                }
        }

        //
        // MAIN program
        //
        int main()
        {
            int CNT = 0;

            while(1)                           // Do forever
            {
                Display(CNT);                  // Display CNT
                wait(1.0);                     // Wait 1 swcond
                CNT++;                         // Increment CNT
                if(CNT == 10)CNT = 0;          // Reset CNT
            }
        }
```

FIG. 7.75,   CONT'D

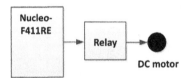

FIG. 7.76   Block diagram of the project.

7.17 PROJECT 14—POWERING LARGE LOADS—DC MOTOR CONTROL

7.17.1 Description

The current capacity of a GPIO pin is limited to 20 mA. There are some applications however where larger currents are needed then the port can supply, such as motors, relays, actuators, etc. In this project, we shall see how a DC motor can be controlled from a GPIO pin of the Nucleo-F411RE development board. This is a very simple project where a relay is used to activate the motor for 10 s, then the motor stops for 5 s, and is then reactivated again for 20 s.

7.17.2 Aim

The aim of this project is to show how loads requiring large currents can be connected to a GPIO pin. Additionally, the project shows how a DC motor can be controlled from a relay.

7.17.3 Block Diagram

The block diagram of the project is shown in Fig. 7.76.

A load requiring large currents can be connected to a microcontroller output pin in one of three ways: using a bipolar transistor, using a MOSFET (Metal Oxide Semiconductor Field Effect Transistor), or using a relay.

Using a Bipolar Transistor

Fig. 7.77 shows how a bipolar transistor can be connected to a load. Here, the transistor is operated as a switch where the load is connected to the collector pin, the emitter pin is connected to ground, and the base is driven from the GPIO pin through a resistor. When a bipolar transistor is operated as a switch, the base current I_B is chosen such that the transistor saturates when a base current is applied. If β is the *minimum* DC current gain of the transistor then the base current should be chosen such that:

$$I_B >= I_L/\beta$$

The base resistor is then chosen using the following formula:

$$R_B = (V_o - 0.7)/I_B$$

where V_o is the output voltage of the GPIO pin when it is at logic 1, and 0.7 is the Base-Emitter DC bias voltage of a bipolar silicon transistor.

As an example, assume that the minimum $\beta = 400$, $V_o = 3.3\,V$, and the load current $I_L = 50\,mA$. The value of the base resistor should then be:

$$I_B \geq 50/400 = 0.125mA$$

Choose $I_B = 0.2\,mA$, then, $R_B = (3.3 - 0.7)/0.2 = 13\,K$, choose 12 K as the nearest physical resistor. It is important to check the data sheets to make sure that the chosen transistor

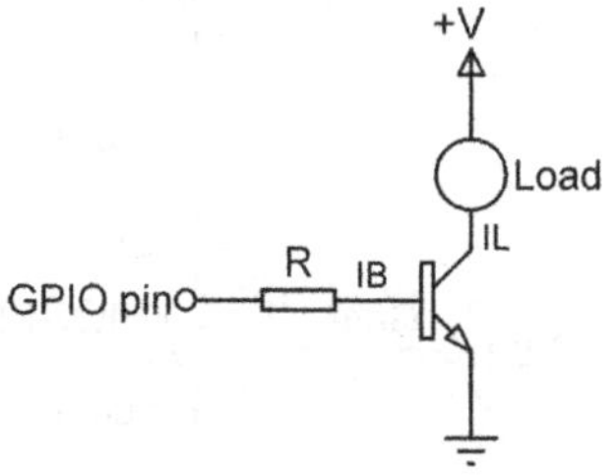

FIG. 7.77   Using a bipolar transistor.

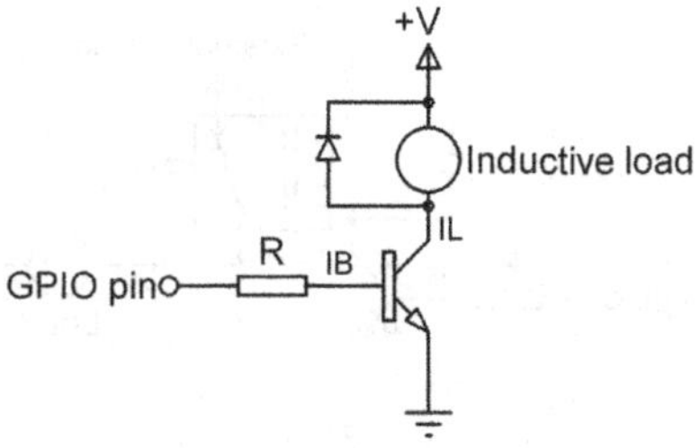

FIG. 7.78   Using a diode to protect the transistor.

maximum collector current is well above the maximum required load current. It may also be necessary to use heatsinks to protect the transistor.

It is recommended to use a *freewheel diode* in parallel with the load, especially with inductive loads to protect the transistor when it is switched off. This is shown in Fig. 7.78.

Using a MOSFET

MOSFETs are used when it is required to power loads requiring larger currents. When choosing a MOSFET we should make sure that the maximum gate-source threshold voltage is lower than the minimum output voltage of the GPIO pin when the pin is at logic 1. Since the output high voltage is +3.3 V minimum, the gate-source threshold voltage should be around 3 V. Fig. 7.79 shows how a MOSFET can be used to switch high current loads. Note here that a freewheel diode will be required as in Fig. 7.78 when driving inductive loads. An example MOSFET that can be used in microcontroller-based applications is the IRL540N. The maximum gate-source threshold voltage of this device is 2 V, and the device can handle load currents up to 36 A.

Using a Relay

Relays can be used when it is required to switch loads requiring very large currents. Relays are normally used as loads in bipolar transistor or in MOSFET circuits. Fig. 7.80 shows how a relay can be used in a bipolar transistor circuit. Note that a freewheel diode will be required in this circuit since the relay winding is inductive.

7.17.4 Circuit Diagram

Fig. 7.81 shows the block diagram of the project. A small 12 V DC motor is connected to a relay through a bipolar transistor switch. The relay is controlled from GPIO pin PC_0 of the Nucleo-F411RE development board.

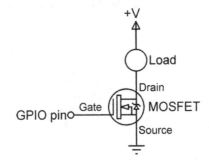

FIG. 7.79   Using a MOSFET.

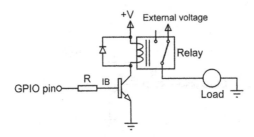

FIG. 7.80   Using a relay.

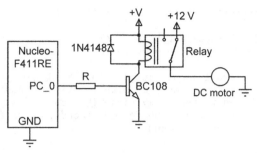

FIG. 7.81   Circuit diagram of the project.

BEGIN
          Configure PC_0 as output
          Turn the motor ON for 10 seconds
          Wait 5 seconds
          Turn the motor ON for 20 seconds
          Stop the motor
END

FIG. 7.82   PDL of the program.

7.17.5 The PDL

The PDL of the program is shown in Fig. 7.82.

7.17.6 Program Listing

The program listing (program: **Relay**) is shown in Fig. 7.83. At the beginning of the program PC_0 is configured as digital output. Then, the relay is activated for 10 s, deactivated for 5 s, and then reactivated for 20 s.

7.17.7 Suggestions for Additional Work

Modify the circuit given in Fig. 7.81 so that a MOSFET is used instead of a bipolar transistor.

7.18 SUMMARY

In this chapter we have learned to develop simple LED-based projects, including the following:

- using the on-board User LED and User button
- using external LEDs and external buttons
- using digital input and output statements
- using the random number generator

```
/***********************************************************************

                    DC MOTOR CONTROL WITH RELAY
                    ===========================

In this program a relay is connected to GPIO pin PC_0 of the nucleo-
F411RE development board through a bipolar transistor. Additionally,
a DC motor is connected to the relay pins. The motor si supplied with
+12V. The project turn the motor ON for 10 seconds, stops for 5 seconds
and then re-starst it for 20 seconds

Author: Dogan Ibrahim
Date  : August 2018
File  : Relay
***********************************************************************/
#include "mbed.h"

DigitalOut Relay(PC_0);        // PC_0 is output

int main()
{
    Relay = 1;                 // Activare relay (START motor)
    wait(10.0);                // Wait 10 seconds
    Relay = 0;                 // de-activate relay (STOP motor)
    wait(5.0);                 // Wait 5 seconds
    Relay = 1;                 // Re-activate the relay (START motor)
    wait(20.0);                // Wait 20 seconds
    Relay = 0;                 // De-activate relay (STOP motor)
}
```

FIG. 7.83   Program listing.

- using an RGB LED
- using a single-digit 7-segment Display
- switching loads requiring large power

7.19 EXERCISES

1. Write a program to flash the User LED 10 times with 0.25s delay between each flashing
2. It is required to connect two external LEDs to a Nucleo-F411RE development board. Draw a possible circuit diagram for this project.
3. Write a program to flash the LEDs alternately with 1s delay in Exercise (2)
4. It is required to design a project having an external button and a 7-segment display. Draw a possible circuit diagram for this project.
5. Write a program to count up on the 7-segment display in Exercise (4) above each time the button is pressed. Assume that the number of presses will be less than 10.
6. It is required to connect an RGB LED to the Nucleo board. Draw the circuit diagram of this project assuming that the PORT C will be used to control the display.
7. Write a program to display the three colors of the RGB as follows: Red for 5 s, Green for 3 s, and Blue for 2 s. Then, display all the colors (i.e., white color) for 10s.

CHAPTER

8

Intermediate Level Projects

8.1 OVERVIEW

In this chapter, we shall be developing more complex projects using the Nucleo-F411RE development board and the Mbed Integrated Development Environment. The projects in this chapter will be described with the same subheadings as in the previous chapter.

8.2 PROJECT 1—TWO-DIGIT MULTIPLEXED 7-SEGMENT LED

8.2.1 Description

A single digit 7-segment LED can only display numbers between 0 and 9. In applications where it is required to display higher numbers, we can multiplex two 7-segment LEDs to display numbers between 0 and 99. In this project, we shall be designing a two-digit display to display the number 25 as an example.

8.2.2 Aim

The aim of this project is to show how two 7-segment LEDs can be multiplexed to display numbers between 0 and 99.

8.2.3 Block Diagram

The block diagram of the project is shown in Fig. 8.1. Notice that in a multiplexed display, the same data are sent to both display segments at the same time, but each digit is enabled separately. Thus, for example, to display number 25, we first send 2 to both digit segments and enable the digit at the left-hand side (MSD, most significant digit) for several milliseconds. Then, number 5 is sent again to both digit segments but this time the digit at the right-hand side (LSD, least significant digit) is enabled for several milliseconds. When this process is repeated, the human eye cannot differentiate that the digits are not ON all the time. This way we can multiplex several digits easily.

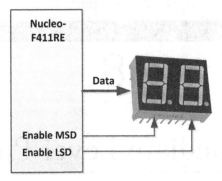

FIG. 8.1　Block diagram of the project.

8.2.4 Circuit Diagram

In this project, the DC56-11EWA two-digit 7-segment display is used. This is a 0.56 in. height common-cathode display having 18 pins. The pin configuration of the display is shown in Fig. 8.2. Pin 1 is at the bottom left-hand side, pin 9 is at the bottom right-hand side, pin 10 is at the top right-hand side, and pin 18 is at the top left-hand side. Notice that the MSD and LSD digits are enabled when pin 14 and pin 13 are logic 0, respectively.

Fig. 8.3 shows the circuit diagram of the project. All the corresponding **a–g** pins of the two digits are connected together and are driven from the lower byte of PORT C (PC_0–PC_6) through 390 ohm current limiting resistors. The two-digit enable pins are controlled using two BC108-type negative-positive-negative (NPN) transistors (any other NPN-type transistor can be used here). GPIO pin PC_8 controls the MSD digit, and pin PC_9 controls the LSD digit. When the base of the transistor is set to logic 1, the corresponding transistor is switched ON and thus its collector becomes at logic 0 which enables the corresponding digit. A digit is disabled if the base of the corresponding transistor is at logic 0.

8.2.5 The Program Design Language

Fig. 8.4 shows the operation of the program as a program design language (PDL).

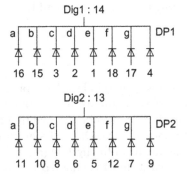

FIG. 8.2　Pin configuration of DC56-11EWA. *Courtesy of Kingbright.*

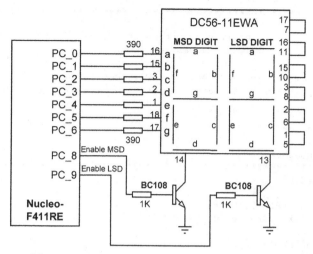

FIG. 8.3   Circuit diagram of the project.

BEGIN
      Group PORT C pins as a single variable
      Define the segment bit patterns
      Configure PC_8 and PC_9 as outputs
      CNT = 25
      DO FOREVER
            Extract the MSD digit of CNT
            Extract the LSD digit of CNT
            Send bit pattern corresponding to MSD to PORT C
            Enable MSD Digit
            Wait for 10ms
            Disable MSD Digit
            Send bit pattern corresponding to LSD to PORT C
            Enable LSD Digit
            Wait for 10ms
            Disable LSD Digit
      ENDDO
   END

FIG. 8.4   Program PDL.

```
/********************************************************************

                    2-DIGIT MULTIPLEXED LED
                    =======================

In this program a 2-digit 7-Segment LED is connected to PORT C of the
Nucleo-F411RE development board. The program displays the number 25 on
the display.

Author: Dogan Ibrahim
Date   : August 2018
File   : SevenSegMux2
*********************************************************************/
#include "mbed.h"

PortOut Segments(PortC, 0xFF);
int LEDS[] = {0x3F,0x06,0x5B,0x4F,0x66,0x6D,0x7D,0x07,0x7F,0x6F};

DigitalOut MSDEnable(PC_8);
DigitalOut LSDEnable(PC_9);

#define Enable 1
#define Disable 0

int main()
{
    int MSDValue, LSDValue, CNT;
    CNT = 25;                              // Number to be displayed
    MSDEnable = Disable;                   // Disable MSD digit
    LSDEnable = Disable;                   // Disable LSD digit

    while(1)                               // Do forever
    {
        MSDValue = CNT / 10;               // MSD of the number
        LSDValue = CNT % 10;               // LSD of the number
        Segments = LEDS[MSDValue];         // Send to PORT C
        MSDEnable = Enable;                // Enable MSD digit
        wait(0.01);                        // Wait 10ms

        MSDEnable = Disable;               // Disable MSD digit
        Segments = LEDS[LSDValue];         // Send to PORT C
        LSDEnable = Enable;                // Enable LSD digit
        wait(0.01);                        // Wait 10ms
        LSDEnable = Disable;               // Disable LSD digit
    }
}
```

FIG. 8.5   Program listing.

8.2.6 Program Listing

The program listing (program: **SevenSegMux2**) is shown in Fig. 8.5. At the beginning of the program, PORT C lower byte bits are grouped together and assigned to variable called **Segments** using the **PortOut** statement. PC_8 and PC_9 bits are defined as **MSDEnable** and **LSDEnable** and are configured as digital outputs. At the beginning of the main program, variable **CNT** is set to 25 and both digits are disabled. The remainder of the program runs in an endless loop established using a **while** statement. Inside this loop, the MSD and the LSD values of the number to be displayed (i.e., CNT) are extracted and saved in variables **MSDValue** and **LSDValue,** respectively. **MSDValue** is then sent to PORT C and **MSDEnable** is set to logic 1 to enable the MSD digit. After 10 ms delay, **MSDEnable** is disabled, **LSDValue** is sent to PORT C, and **LSDEnable** is enabled. This digit is disabled after 10 ms and the above process repeats forever until stopped by the user.

8.2.7 Modified Program

In the program in Fig. 8.5, a number less than 10 is displayed with a leading 0. For example, number 5 is displayed as 05. We can disable the MSD digit if it is 0. This is shown in the program in Fig. 8.6 (program: **SevenSegMux3**) where number 5 is displayed on the LED.

8.2.8 Suggestions for Additional Work

In Fig. 8.5, modify the digit enable time and see its effects on the display.

Modify the program given in Fig. 8.5 to display the letters HE and then LO continuously with 1 s delay between each display.

8.3 PROJECT 2—FOUR-DIGIT MULTIPLEXED 7-SEGMENT LED

8.3.1 Description

A two-digit 7-segment LED can only display numbers between 0 and 99. In applications where it is required to display higher numbers, we can multiplex three or higher 7-segment LEDs together. In this project, a 7-segment display is formed by multiplexing four LEDs so that numbers 0–9999 can be displayed. In this project, the number 3579 is displayed as an example.

8.3.2 Aim

The aim of this project is to show how four 7-segment LEDs can be multiplexed to display numbers between 0 and 9999.

8.3.3 Block Diagram

The block diagram of the project is shown in Fig. 8.7. As with a two-digit display, the same data are sent to all four display segments at the same time, but each digit is enabled for a short time. The human eye sees that all the four digits are enabled at all times.

8.3.4 Circuit Diagram

Although there are four-digit 7-segment LED packages, in this project 2 × DC56-11EWA-type two-digit LEDs are used to form a four-digit LED as shown in the block diagram in

```
/*********************************************************************

                    2-DIGIT MULTIPLEXED LED
                    =======================

In this program a 2-digit 7-Segment LED is connected to PORT C of the
Nucleo-F411RE development board. The program displays the number 5 on
the display.

In this modified program leading 0 is disabled.

Author: Dogan Ibrahim
Date   : August 2018
File   : SevenSegMux3
*********************************************************************/
#include "mbed.h"

PortOut Segments(PortC, 0xFF);
int LEDS[] = {0x3F,0x06,0x5B,0x4F,0x66,0x6D,0x7D,0x07,0x7F,0x6F};

DigitalOut MSDEnable(PC_8);
DigitalOut LSDEnable(PC_9);

#define Enable 1
#define Disable 0

int main()
{
    int MSDValue, LSDValue, CNT;
    CNT = 5;                            // Number to be displayed
    MSDEnable = Disable;                // Disable MSD digit
    LSDEnable = Disable;                // Disable LSD digit

    while(1)                            // Do forever
    {
        MSDValue = CNT / 10;            // MSD of the number
        LSDValue = CNT % 10;            // LSD of the number
        if(MSDValue != 0)               // If MSD value is not 0
        {
            Segments = LEDS[MSDValue];  // Send to PORT C
            MSDEnable = Enable;         // Enable MSD digit
            wait(0.01);                 // Wait 10ms
            MSDEnable = Disable;        // Disable MSD digit
        }

        Segments = LEDS[LSDValue];      // Send to PORT C
        LSDEnable = Enable;             // Enable LSD digit
        wait(0.01);                     // Wait 10ms
        LSDEnable = Disable;            // Disable LSD digit
    }
}
```

FIG. 8.6   Modified program.

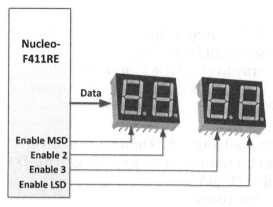

FIG. 8.7   Block diagram of the project.

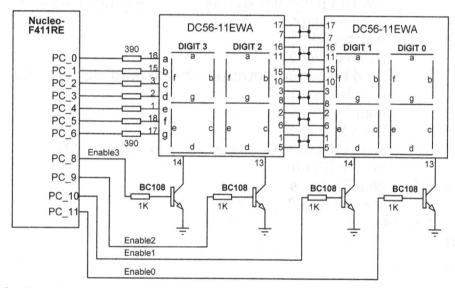

FIG. 8.8   Circuit diagram of the project.

Fig. 8.7. Fig. 8.8 shows the project circuit diagram. All the corresponding **a–g** pins of the four digits are connected together and are driven from the lower byte of PORT C (PC_0–PC_6) through 390 ohm current limiting resistors. The four-digit enable pins are controlled using two BC108-type NPN transistors (any other NPN-type transistor can be used here). GPIO pin PC_8, PC_9, PC_10, and PC_11 control the digits as shown in Fig. 8.8. When the base of the transistor is set to logic 1, the corresponding transistor is switched ON and thus its collector becomes at logic 0 which enables the corresponding digit. A digit is disabled if the base of the corresponding transistor is at logic 0.

8.3.5 The PDL

Fig. 8.9 shows the operation of the program as a PDL.

BEGIN
　　Group PORT C pins as a single variable
　　Define the segment bit patterns
　　Create an array to store the four digits values
　　Configure PC_8,PC_9,PC_10,PC11as outputs
　　CNT = 3579
　　DO FOREVER
　　　　Extract all four digits and store in an array
　　　　Send bit pattern corresponding to MSD to PORT C
　　　　Enable Digit3
　　　　Wait for 10ms
　　　　DisableMSD Digit
　　　　Send bit pattern corresponding to second digitto PORT C
　　　　Enable Digit2
　　　　Wait for 10ms
　　　　Disable Digit2
　　　　Send bit pattern corresponding to third digit to PORT C
　　　　Enable Digit 1
　　　　Wait for 10ms
　　　　Disable Digit1
　　　　Send bit pattern corresponding to LSD to PORT C
　　　　Enable Digit0
　　　　Wait for 10ms
　　　　Disable Digit0
　　ENDDO
END

FIG. 8.9　Program PDL.

8.3.6 Program Listing

The program listing (program: **SevenSegMux4**) is shown in Fig. 8.10. At the beginning of the program, PORT C lower byte bits are grouped together and assigned to variable called **Segments** using the **PortOut** statement. PC_8, PC_9, PC_10, and PC_11 bits are defined as **Enable3**, **Enable2**, **Enable1**, and **Enable0** and are configured as digital outputs. At the beginning of the main program, variable **CNT** is set to 3579 and all four digits are disabled. The remainder of the program runs in an endless loop established using a **while** statement. Inside this loop, the digit values of the number to be displayed (i.e., CNT) are extracted and stored in an array called **Digits**, where **Digits[0]** corresponds to **Digit 0** value, and **Digits[3]** corresponds to **Digit 3** value. The digit values are then sent to PORT C and the corresponding digits are enabled for 5ms as in the previous project. The above process is repeated forever until stopped by the user.

```
/***********************************************************************

                    4-DIGIT MULTIPLEXED LED
                    =======================

In this program two 2-digit 7-Segment LEDs are connected to PORT C of the
Nucleo-F411RE development board to form a 4-digit 7-Segment display. The
program displays the number 3579 on the display.


Author: Dogan Ibrahim
Date  : August 2018
File  : SevenSegMux4
***********************************************************************/
#include "mbed.h"

PortOut Segments(PortC, 0xFF);
int LEDS[] = {0x3F,0x06,0x5B,0x4F,0x66,0x6D,0x7D,0x07,0x7F,0x6F};

DigitalOut Enable3(PC_8);
DigitalOut Enable2(PC_9);
DigitalOut Enable1(PC_10);
DigitalOut Enable0(PC_11);

#define Enable 1
#define Disable 0

int main()
{
    int Y, W, CNT = 3579;
    int Digits[4];                          // Array to store digits

    Enable3 = Disable;                      // Disable Digit 3
    Enable2 = Disable;                      // Disable Digit 2
    Enable1 = Disable;                      // Disable Digit 3
    Enable0 = Disable;                      // Disable Digit 3

//
// Extract digits of CNT into Digits[]. Digits[3] holds the MSD digit
// and Digits[0] holds the LSD digit
//
    Digits[3] = CNT / 1000;
    Y = CNT - 1000*Digits[3];
    Digits[2] = Y / 100;
    W = Y - 100*Digits[2];
    Digits[1] = W /10;
    Digits[0] = W % 10;

    while(1)
```

FIG. 8.10 Program listing.

(Continued)

```
{
        Segments = LEDS[Digits[3]];       // Send to PORT C
        Enable3 = Enable;                 // Enable Digit 3
        wait(0.005);                      // Wait 5ms
        Enable3 = Disable;                // Disable Digit 3

        Segments = LEDS[Digits[2]];       // Send to PORT C
        Enable2 = Enable;                 // Enable Digit 2
        wait(0.005);                      // Wait 5ms
        Enable2 = Disable;                // Disable Digit 2

        Segments = LEDS[Digits[1]];       // Send to PORT C
        Enable1 = Enable;                 // Enable Digit 1
        wait(0.005);                      // Wait 5ms
        Enable1 = Disable;                // Disable Digit 1

        Segments = LEDS[Digits[0]];       // Send to PORT C
        Enable0 = Enable;                 // Enable Digit 0
        wait(0.005);                      // Wait 5ms
        Enable0 = Disable;                // Disable Digit 0
    }
}
```

FIG. 8.10,   CONT'D

There are several algorithms for extracting the digits of an integer number. In this project, the number to be displayed cannot be greater than four digits long. Assuming that the number CNT is, for example, 2356, the digits are extracted as follows:

Calculations	Digits Extracted
Digits[3]=CNT/1000;	Digits[3]=2 (MSD)
Y=CNT−1000*Digits[3];	Y=356
Digits[2]=Y/100;	Digits[2]=3
W=Y−100*Digits[2];	W=56
Digits[1]=W/10;	Digits[1]=5
Digits[0]=W% 10;	Digits[0]=6 (LSD)

8.3.7 Modified Program

The program in Fig. 8.10 displays the leading zeroes. For example, number 10 is displayed as 0010. We can remove the leading zeroes by disabling their digit. For example, when displaying "0010," we can disable the **Enable3** and **Enable2** digits so that the display shows "10" and not "0010." The program given in Fig. 8.11 (program: **SevenSegMux5**) shows how the leading zeros can be disabled.

8.3.8 Suggestions for Additional Work

Modify the program as shown in Fig. 8.10 to display the letters "HELO."

```
/**************************************************************************
                        4-DIGIT MULTIPLEXED LED
                        =======================

In this program two 2-digit 7-Segment LEDs are connected to PORT C of the
Nucleo-F411RE development board to form a 4-digit 7-Segment display. The
program displays the number 5 on the display.

In this modified program the leading zeroes are disabled so that for
example number 5 is displayed as "   5" and not as "0005"

Author: Dogan Ibrahim
Date   : August 2018
File   : SevenSegMux5
**************************************************************************/
#include "mbed.h"

PortOut Segments(PortC, 0xFF);
int LEDS[] = {0x3F,0x06,0x5B,0x4F,0x66,0x6D,0x7D,0x07,0x7F,0x6F};

DigitalOut Enable3(PC_8);
DigitalOut Enable2(PC_9);
DigitalOut Enable1(PC_10);
DigitalOut Enable0(PC_11);

#define Enable 1
#define Disable 0

int main()
{
    int W, Y, CNT;                              // Number to be displayed
    int Digits[4];                              // Array to store digits

    Enable3 = Disable;                          // Disable Digit 3
    Enable2 = Disable;                          // Disable Digit 2
    Enable1 = Disable;                          // Disable Digit 1
    Enable0 = Disable;                          // Disable Digit 0
//
// Extract digits of CNT into Digits[]. Digits[3] holds the MSD digit
// and Digits[0] holds the LSD digit
//
    CNT = 5;                                    // Number to be displayed
    Digits[3] = CNT / 1000;
    Y = CNT - 1000*Digits[3];
    Digits[2] = Y / 100;
    W = Y - 100*Digits[2];
    Digits[1] = W /10;
    Digits[0] = W % 10;

    while(1)                                    // Do forever
```

FIG. 8.11 Modified program.

(Continued)

```
    {
        if(Digits[3] != 0)                          // If MSD Digit non zero
        {
            Segments = LEDS[Digits[3]];             // Send to PORT C
            Enable3 = Enable;                       // Enable Digit 3
            wait(0.005);                            // Wait 5ms
            Enable3 = Disable;                      // Disable Digit 3
        }

        if(Digits[2] != 0 || (Digits[2] == 0 && Digits[3] != 0))
        {
            Segments = LEDS[Digits[2]];             // Send to PORT C
            Enable2 = Enable;                       // Enable Digit 2
            wait(0.005);                            // Wait 5ms
            Enable2 = Disable;                      // Disable Digit 2
        }

        if(Digits[1] != 0 || (Digits[3] != 0 || Digits[2] != 0))
        {
            Segments = LEDS[Digits[1]];             // Send to PORT C
            Enable1 = Enable;                       // Enable Digit 1
            wait(0.005);                            // Wait 5ms
            Enable1 = Disable;                      // Disable Digit 1
        }

        Segments = LEDS[Digits[0]];                 // Send to PORT C
        Enable0 = Enable;                           // Enable Digit 0
        wait(0.005);                                // Wait 5ms
        Enable0 = Disable;                          // Disable Digit 0
    }
}
```

FIG. 8.11, CONT'D

8.4 POLLING AND INTERRUPTS

In many real-time applications, we may want to respond to external events as soon as they happen. But this seems not to be possible while the CPU is executing some code. What we basically want is to interrupt the CPU so that it stops what it has been doing and starts to execute the code in our real-time routine. After completing the real-time routine, the CPU should return and carry on what it has been doing before being interrupted. The concept that stops the CPU and starts executing the real-time code is called *Interrupt*.

Basically, there are two ways that the CPU can respond to external events: *polling* and *interrupts*.

8.4.1 Polling

In polling, the CPU checks the occurrence of an external event in a loop while executing some other code. Fig. 8.12 shows a flow diagram where the CPU is carrying out some

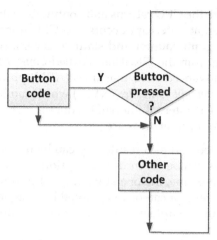

FIG. 8.12   Program using polling.

operations and at the same time, it is required to respond when an external button is pressed. It is clear from this figure that the CPU is not very responsive to the button event and it can only check the state of the button when it completes executing the other code.

Polling has the disadvantage that the CPU cannot respond to external events quickly during a polling routine. Also, if there is an error in the other code and the code happens to stay in an indefinite loop, then the CPU will never respond to the external events.

8.4.2 Interrupts

Interrupts are very important parts of all microcontrollers as they allow the microcontroller to respond to external events very quickly. As shown in Fig. 8.13, when an interrupt occurs due to an external or an internal event (e.g., the timer), the CPU stops executing the current instruction and jumps to execute the code named the *interrupt service routine* (ISR). After

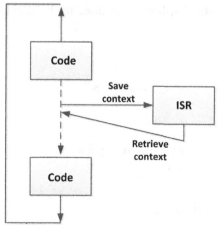

FIG. 8.13   The interrupt process.

completing executing the ISR, the CPU returns and continues to the point where it left before the interrupt occurred. When an interrupt occurs, the CPU normally saves the contents of some registers (e.g., the program counter and status registers) so that it can retrieve them and continue after returning from the ISR. This is called *Context Saving*.

On the Cortex-M-based advanced RISC machines (ARM) processors interrupts are managed by the *nested vectored interrupt controller* (NVIC) which can handle and manage several hundreds of external as well as internal interrupt requests.

Some important properties of interrupts are:

- Interrupts can be prioritized and as a result they can be nested. What this means is that different interrupt sources can be given different priority levels so that when multiple interrupts occur at the same time, the one at the highest priority is handled and executed first. Also, an executing interrupt can be interrupted by a higher-priority interrupt. When the higher-priority interrupt completes executing its ISR, control returns to the lower priority pending interrupt.
- Interrupts can be masked so that, for example, unwanted interrupts from interrupt sources can be disabled. A disabled interrupt source cannot interrupt the CPU.
- ISR routines start from specific locations in memory, called the interrupt *vector* addresses.

8.5 PROJECT 3—FOUR-DIGIT 7-SEGMENT LED COUNTER

8.5.1 Description

The problem of using a multiplexed 7-segment LED in a microcontroller is that the CPU has to refresh the display at regular intervals (e.g., at every 5 ms) and as a result of this, the CPU cannot do other tasks. One way round this problem is to use a timer ISR at 5 ms intervals and refresh the display inside this timer interrupt routine. In this project, the display will count up every second from 0 to 9999.

8.5.2 Aim

The aim of this project is to show how the 7-segment LED display code can be inserted inside to a timer ISR so that the display is refreshed in the background. This way, the CPU is free to do other tasks.

8.5.3 Block Diagram

The block diagram of the project is as in Fig. 8.7.

8.5.4 Circuit Diagram

The circuit diagram of the project is as in Fig. 8.8.

8.5.5 The PDL

The PDL of the program is shown in Fig. 8.14.

BEGIN/MAIN
    Group PORT C pins as a single variable
    Define the segment bit patterns
    Create an array to store the four digits values
    Configure PC_8,PC_9,PC_10,PC11 as outputs
    Create a Ticker variable and attach to function Refresh
    CNT = 0
    Flag = 3
    DO FOREVER
        Increment CNT
        Wait 1 second
    ENDDO
END/MAIN

BEGIN/Refresh
    DO FOREVER
        Extract all four digits and store in an array
        IF Flag = 3 THEN
            Flag = 2
            Disable Digit 0
            Send bit pattern corresponding to Digit 3 to PORT C
            Enable Digit 3
        ELSE IF Flag = 2 THEN
            Flag = 1
            Disable Digit 3
            Send bit pattern corresponding to Digit 2 to PORT C
            Enable Digit 2
        ELSE IF Flag = 1 THEN
            Flag = 0
            Disable Digit 2
            Send bit pattern corresponding to Digit 1 to PORT C
            Enable Digit 1
        ENDIF
        IF Flag = 0 THEN
            Flag = 3
            Disable Digit 1
            Send bit pattern corresponding to Digit 0 to PORT C
            Enable Digit 0
    ENDDO
END/Refresh

FIG. 8.14   Program PDL.

Mbed supports three types of timers [in addition to the real-time clock (RTC)]: **Timer**, **Timeout**, and **Ticker**. In this section, we shall be looking at the details of Ticker function which can be used to generate timer interrupts at regular intervals. We shall be looking at the use of other timer functions in later sections of this chapter.

Using the Ticker

The ticker function is used to call an ISR periodically at regular intervals. Table 8.1 presents the various Ticker functions. The steps to use the Ticker function are as follows:

- Define a Ticket-type variable
- Attach the Ticker variable to a function (this is your ISR) and define its period
- The ISR will be called periodically at the specified times

8.5.6 Program Listing

The program listing (program: **SevenSegMux6**) is given in Fig. 8.15. At the beginning of the program, **tim** is defined of type **Ticker**, lower byte of PORT C is grouped together into a variable called **Segments**. Array **Digits** is declared at the beginning of the program so that it is a global integer array. All the digits are disabled, and function called **Refresh** has been attached to the **Ticker** with a period of 5 ms so that the function is called periodically at every 5 ms. The main program runs in a loop, increments variable **CNT**, and waits for 1 s. Function **Refresh** implements the display functions. Here, variable **Flag** is used to decide which digit should be refreshed. If **Flag** is 3 then digit 3 is refreshed, if it is 2 then digit 2 is refreshed, and so on. Notice that the **wait** statements have been removed from the program since the **Refresh** function is called at every 5 ms. Also, although digit 3 is enabled in digit 3 refresh cycle, it is disabled in digit 2 refresh cycle. This is done for the other digits as well.

8.6 PROJECT 4—FOUR-DIGIT 7-SEGMENT LED EVENT COUNTER

8.6.1 Description

In this project, the Ticker timer interrupts are used to refresh the four-digit 7-segment display as in the previous project. The program counts up by one every time the on-board User button is pressed, therefore simulating the occurrence of external events.

TABLE 8.1   Ticker Functions

Timeout Function	Description
attach	Attach a user function to a Ticker, specifying the repeat interval in seconds
attach_us	Attach a user function to a Ticker, specifying the repeat interval in microseconds
detach	Detach the function from Ticker

```
/****************************************************************************
                   4-DIGIT MULTIPLEXED LED COUNTER
                   ===============================

In this program two 2-digit 7-Segment LEDs are connected to PORT C of the
Nucleo-F411RE development board to form a 4-digit 7-Segment display. The
program counts up every second starting from 0. Leading zeroes are disabled
in this program. The display routine runs insid ethe timer interrupt service
routine called Refresh, which is called automatically every 5ms.


Author: Dogan Ibrahim
Date  : August 2018
File  : SevenSegMux6
*****************************************************************************/
#include "mbed.h"
Ticker tim;

PortOut Segments(PortC, 0xFF);
int LEDS[] = {0x3F,0x06,0x5B,0x4F,0x66,0x6D,0x7D,0x07,0x7F,0x6F};
int W, Y, CNT = 0, Flag = 3;
int Digits[4];

//
// Digit Enable bits
//
DigitalOut Enable3(PC_8);
DigitalOut Enable2(PC_9);
DigitalOut Enable1(PC_10);
DigitalOut Enable0(PC_11);

#define Enable 1
#define Disable 0

//
// Thsi is the Interrupt Service Routine (ISR) which is called at every
// 5ms by Ticker
//
void Refresh()
{
//
// Extract digits of CNT into Digits[]. Digits[3] holds the MSD digit
// and Digits[0] holds the LSD digit
//
    Digits[3] = CNT / 1000;
    Y = CNT - 1000*Digits[3];
    Digits[2] = Y / 100;
    W = Y - 100*Digits[2];
    Digits[1] = W /10;
    Digits[0] = W % 10;

    if(Flag == 3)                                // If to refresh Digit 3
    {
```

FIG. 8.15   Program listing.

(Continued)

```
            Enable0 = Disable;
            Flag = 2;
            if(Digits[3] != 0)                    // If MSD Digit non zero
            {
                Segments = LEDS[Digits[3]];        // Send to PORT C
                Enable3 = Enable;                  // Enable Digit 3
            }
    }
    else if(Flag == 2)                            // If to refresh Digit 2
    {
        Enable3 = Disable;
        Flag = 1;
        if(Digits[2] != 0 || (Digits[2] == 0 && Digits[3] != 0))
        {
            Segments = LEDS[Digits[2]];            // Send to PORT C
            Enable2 = Enable;                      // Enable Digit 2
        }
    }
    else if(Flag == 1)                            // If to refresh Digit 1
    {
        Enable2 = Disable;
        Flag = 0;
        if(Digits[1] != 0 || (Digits[3] != 0 || Digits[2] != 0))
        {
            Segments = LEDS[Digits[1]];            // Send to PORT C
            Enable1 = Enable;                      // Enable Digit 1
        }
    }

    else if(Flag == 0)                            // If to refresh Digit 0
    {
        Enable1 = Disable;
        Flag = 3;
        Segments = LEDS[Digits[0]];                // Send to PORT C
        Enable0 = Enable;                          // Enable Digit 0
    }
}


int main()
{

    Enable3 = Disable;                            // Disable Digit 3
    Enable2 = Disable;                            // Disable Digit 2
    Enable1 = Disable;                            // Disable Digit 1
    Enable0 = Disable;                            // Disable Digit 0

    tim.attach(&Refresh, 0.005);

    while(1)                                       // Do forever
    {
        CNT++;                                     // Increment CNT
        wait(1.0);                                 // Wait 1 second
    }
}
```

FIG. 8.15,   CONT'D

8.6.2 Aim

The aim of this project is to show how an event counter can be designed and the events displayed on a four-digit multiplexed 7-segment LED display. It is assumed that pressing the User button creates an external event.

8.6.3 Block Diagram

The block diagram of the project is as in Fig. 8.16.

8.6.4 Circuit Diagram

The circuit diagram of the project is as in Fig. 8.8. In addition, the onboard User button is used to create external events.

8.6.5 The PDL

The PDL of the program is very similar to the PDL given in Fig. 8.14, but here the count is incremented when the User button is pressed.

8.6.6 Program Listing

Fig. 8.17 shows the program listing (program: **EventCounter**). The program listing is very similar to the one given in Fig. 8.15. In this program, the User button is configured as digital input. In this program, the count (CNT) is incremented by one inside the main program every time the button is pressed. The delay has been removed from the main program. The ISR (function **Refresh**) refreshes the display at every 5ms.

8.6.7 Suggestions for Additional Work

Modify the program given in Fig. 8.17 so that the count starts from 1000 and counts done by one every time the button is pressed.

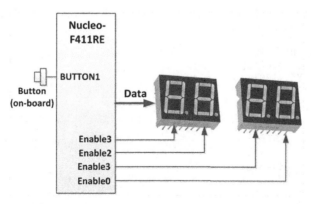

FIG. 8.16   Block diagram of the project.

```
/*****************************************************************************

                  4-DIGIT MULTIPLEXED LED EVENT COUNTER
                  =======================================

In this program two 2-digit 7-Segment LEDs are connected to PORT C of the
Nucleo-F411RE development board to form a 4-digit 7-Segment display. The
program counts up by one each time the User Button is pressed. The button
simulates the occurence of external events.

Author: Dogan Ibrahim
Date   : August 2018
File   : EventCounter
*****************************************************************************/
#include "mbed.h"
Ticker tim;

PortOut Segments(PortC, 0xFF);
int LEDS[] = {0x3F,0x06,0x5B,0x4F,0x66,0x6D,0x7D,0x07,0x7F,0x6F};
DigitalIn button(BUTTON1);
int W, Y, CNT = 0, Flag = 3;
int Digits[4];

//
// Digit Enable bits
//
DigitalOut Enable3(PC_8);
DigitalOut Enable2(PC_9);
DigitalOut Enable1(PC_10);
DigitalOut Enable0(PC_11);

#define Enable 1
#define Disable 0

//
// Thsi is the Interrupt Service Routine (ISR) which is called at every
// 5ms by Ticker
//
void Refresh()
{
//
// Extract digits of CNT into Digits[]. Digits[3] holds the MSD digit
// and Digits[0] holds the LSD digit
//
    Digits[3] = CNT / 1000;
    Y = CNT - 1000*Digits[3];
    Digits[2] = Y / 100;
    W = Y - 100*Digits[2];
    Digits[1] = W /10;
    Digits[0] = W % 10;

    if(Flag == 3)                                    // If to refresh Digit 3
    {
        Enable0 = Disable;
        Flag = 2;
```

FIG. 8.17   Program listing.

(Continued)

```
        if(Digits[3] != 0)                      // If MSD Digit non zero
        {
            Segments = LEDS[Digits[3]];          // Send to PORT C
            Enable3 = Enable;                    // Enable Digit 3
        }
    }
    else if(Flag == 2)                          // If to refresh Digit 2
    {
        Enable3 = Disable;
        Flag = 1;
        if(Digits[2] != 0 || (Digits[2] == 0 && Digits[3] != 0))
        {
            Segments = LEDS[Digits[2]];          // Send to PORT C
            Enable2 = Enable;                    // Enable Digit 2
        }
    }
    else if(Flag == 1)                          // If to refresh Digit 1
    {
        Enable2 = Disable;
        Flag = 0;
        if(Digits[1] != 0 || (Digits[3] != 0 || Digits[2] != 0))
        {
            Segments = LEDS[Digits[1]];          // Send to PORT C
            Enable1 = Enable;                    // Enable Digit 1
        }
    }

    else if(Flag == 0)                          // If to refresh Digit 0
    {
        Enable1 = Disable;
        Flag = 3;
        Segments = LEDS[Digits[0]];              // Send to PORT C
        Enable0 = Enable;                        // Enable Digit 0
    }
}

int main()
{

    Enable3 = Disable;                          // Disable Digit 3
    Enable2 = Disable;                          // Disable Digit 2
    Enable1 = Disable;                          // Disable Digit 1
    Enable0 = Disable;                          // Disable Digit 0

    tim.attach(&Refresh, 0.005);

    while(1)                                    // Do forever
    {
        if(button == 0)                         // If button is pressed
        {
            CNT++;                              // Increment CNT
            while(button == 0);                 // Wait until button released
        }
    }
}
```

FIG. 8.17, CONT'D

8.7 PROJECT 5—7-SEGMENT LED DICE

8.7.1 Description

This is a dice project. When the User button is pressed, two dice numbers (between 1 and 6) are displayed on the four-digit 7-segment LED. Normally, a 0 is displayed which means that the project is ready and the user can press the button. The two generated numbers are displayed such that digit 3 and digit 1 are disabled. For example, the numbers 2 and 4 are displayed as "x 2x4," where x is a blank. The numbers are displayed for 5s. After this time, a 0 is displayed to indicate that the system is ready and the user can press the button again to generate a set of new dice numbers.

8.7.2 Aim

The aim of this project is to show how an electronic dice can be designed using the Nucleo-F411RE development board and a four-digit 7-segment display.

8.7.3 Block Diagram

The block diagram of the project is as in Fig. 8.16.

8.7.4 Circuit Diagram

The circuit diagram of the project is as in Fig. 8.8. Additionally, the onboard User button is used to create external events.

8.7.5 The PDL

The PDL of the program is shown in Fig. 8.18.

8.7.6 Program Listing

Fig. 8.19 shows the program listing (program: **DICE**). The program listing is very similar to the one given in Fig. 8.15. In this program, the User button is configured as digital input. **tim** is defined of type **Ticker**, lower byte of PORT C is grouped together into a variable called **Segments**. Array **Digits** is declared at the beginning of the program so that it is a global integer array. All the digits are disabled, and function called **Refresh** has been attached to the **Ticker** with a period of 5 ms so that the function is called periodically at every 5 ms. The main program runs in a loop. Initially, variable **Dice** is set to 0 so that 0 is displayed to indicate that the system is ready for the user to press the User button to generate the dice numbers. The generated numbers are stored in variables **Temp1** and **Temp2**. Variable **Dice** is then set to: **Dice = Temp1 * 100 + Temp2** so that the two dice numbers are displayed at digit positions 2 and 0. The numbers are displayed for 5 s, and after this time variable **Dice** is set to 0 to indicate that the system is ready again.

BEGIN/MAIN
        Group PORT C pins as a single variable
        Define the segment bit patterns
        Create an array to store the four digits values
        Configure PC_8,PC_9,PC_10,PC11 as outputs
        Create a Ticker variable and attach to function Refresh
        Dice = 0
        Flag = 3
        DO FOREVER
                Generate a random number 1 to 6 in Temp1
                Generate a second random number 1 to 6 in Temp2
                Dice = Temp1 * 100 + Temp2
                Wait 5 seconds
                Dice = 0
        ENDDO
END/MAIN

BEGIN/Refresh
        DO FOREVER
                Extract all four digits and store in an array
                IF Flag = 3 THEN
                        Flag = 2
                        Disable Digit 0
                        Send bit pattern corresponding to Digit 3 to PORT C
                        Enable Digit 3
                ELSE IF Flag = 2 THEN
                        Flag = 1
                        Disable Digit 3
                        Send bit pattern corresponding to Digit 2 to PORT C
                        Enable Digit 2
                ELSE IF Flag = 1 THEN
                        Flag = 0
                        Disable Digit 2
                ENDIF
                IF Flag = 0 THEN
                        Flag = 3
                        Disable Digit 1
                        Send bit pattern corresponding to Digit 0 to PORT C
                        Enable Digit 0
        ENDDO
END/Refresh

FIG. 8.18 Program PDL.

```
/************************************************************************
                              DICE
                              ====

In this program two 2-digit 7-Segment LEDs are connected to PORT C of the
Nucleo-F411RE development board to form a 4-digit 7-Segment display. The
program simulates two dice. Each time the User Button is pressed two dice
numbers are displayed between 1 and 6 on the 4-digit multiplexed LED. The
program is ready for the user to press the button when the display shows 0.
Notice that Digit 1 is disabled so that the displayed dice numbers are as
follows: For example if the numbers are 2,4, they are displayed as " 2 4"
and not as " 204"


Author: Dogan Ibrahim
Date   : August 2018
File   : DICE
*************************************************************************/
#include "mbed.h"
Ticker tim;

PortOut Segments(PortC, 0xFF);
int LEDS[] = {0x3F,0x06,0x5B,0x4F,0x66,0x6D,0x7D,0x07,0x7F,0x6F};
DigitalIn button(BUTTON1);
int W, Y, Dice = 0, Flag = 3;
int Digits[4];

//
// Digit Enable bits
//
DigitalOut Enable3(PC_8);
DigitalOut Enable2(PC_9);
DigitalOut Enable1(PC_10);
DigitalOut Enable0(PC_11);

#define Enable 1
#define Disable 0

//
// Thsi is the Interrupt Service Routine (ISR) which is called at every
// 5ms by Ticker
//
void Refresh()
{
//
// Extract digits of CNT into Digits[]. Digits[3] holds the MSD digit
// and Digits[0] holds the LSD digit
//
    Digits[3] = Dice / 1000;
    Y = Dice - 1000*Digits[3];
    Digits[2] = Y / 100;
    W = Y - 100*Digits[2];
    Digits[1] = W /10;
    Digits[0] = W % 10;
```

FIG. 8.19   Program listing.

(Continued)

```
        if(Flag == 3)                          // If to refresh Digit 3
        {
            Enable0 = Disable;
            Flag = 2;
            if(Digits[3] != 0)                  // If MSD Digit non zero
            {
                Segments = LEDS[Digits[3]];     // Send to PORT C
                Enable3 = Enable;               // Enable Digit 3
            }
        }
        else if(Flag == 2)                      // If to refresh Digit 2
        {
            Enable3 = Disable;
            Flag = 1;
            if(Digits[2] != 0 || (Digits[2] == 0 && Digits[3] != 0))
            {
                Segments = LEDS[Digits[2]];     // Send to PORT C
                Enable2 = Enable;               // Enable Digit 2
            }
        }
        else if(Flag == 1)                      // If to refresh Digit 1
        {
            Enable2 = Disable;
            Flag = 0;
            if(Digits[1] != 0 || (Digits[3] != 0 || Digits[2] != 0))
            {
                Segments = LEDS[Digits[1]];     // Send to PORT C
                //Enable1 = Enable;             // Disable Digit 1
            }
        }

        else if(Flag == 0)                      // If to refresh Digit 0
        {
            Enable1 = Disable;
            Flag = 3;
            Segments = LEDS[Digits[0]];         // Send to PORT C
            Enable0 = Enable;                   // Enable Digit 0
        }
    }

int main()
{
    int Temp1, Temp2;

    Enable3 = Disable;                          // Disable Digit 3
    Enable2 = Disable;                          // Disable Digit 2
    Enable1 = Disable;                          // Disable Digit 1
    Enable0 = Disable;                          // Disable Digit 0

    tim.attach(&Refresh, 0.005);

    Dice = 0;                                   // Ready to start
    while(1)                                    // Do forever
    {
        if(button == 0)                         // If button is pressed
        {
            Temp1 = rand() % 6 +1;              // First number 1-6
            Temp2 = rand() % 6 + 1;             // Second number 1-6
            Dice = Temp1*100 + Temp2;           // Dice number
            wait(5.0);                          // Wait 5 seconds
            Dice = 0;                           // Ready to start
        }
    }
}
```

FIG. 8.19, CONT'D

8.8 PC SERIAL INTERFACE

The Nucleo-F411RE development board is connected to a PC through its mini USB port. The development board receives its power from the PC, also user programs are uploaded to the development board through the USB interface. The USB interface can also be used to communicate with the PC and exchange data. This feature is useful for several reasons:

- The PC acts like a large reliable display with a keyboard. As a result, the output of the programs developed on the Nucleo board can be displayed on the PC screen. This can be very useful during the program development. For example, the output of a temperature sensor can be displayed on the PC screen while a temperature controller system is being developed.
- The developed programs can be debugged easily by placing messages at various points inside the programs. This speeds up the development cycle and helps to remove any errors from the program.
- The PC keyboard can be used to send messages to the development board. Additionally, data can be received from the PC keyboard for testing purposes. Again, these can be very useful during the development of complex programs.

A terminal emulation program is required on the PC in order to communicate with the development board. There are several freely available terminal emulation software on the PC, such as **HyperTerm, Tera Term**, **Putty**, etc. In this book, we shall be using the Putty terminal emulation software.

8.8.1 Using the Putty

Putty is a popular terminal emulator program that can be downloaded from the following site free of charge:

https://www.putty.org

Activate Putty by clicking on it (**Putty.exe**) after it has been downloaded. Putty can be used for Raw, Telnet, SSH, Rlogin, or Serial interface. Here, we shall be using the Serial option. Before using the Putty, we have to know the serial port number used to communicate with the Nucleo board. This can be found as follows (for Windows 7):

- Plug-in your Nucleo-F411RE development board to the USB port of your PC
- Click **Start** and then **Control Panel** and then click on **System**
- Click on **Device Manager** and then on **Ports**
- You should see the port number listed as a **Virtual COM Port** as shown in Fig. 8.20. In this example, the port number is **COM51**
- Close the **Device Manager** tabs

Now that we know the serial port number we can start using the Putty. The steps are given below (Fig. 8.21):

- Activate Putty
- Select Serial and enter COM51 for the Serial line, and 9600 for the Speed as shown in Fig. 8.22. By default, the Nucleo-F411RE serial port settings are: speed is set to 9600, data bits to 8, no parity, and 1 stop bit
- Click **Open** to open the Putty screen

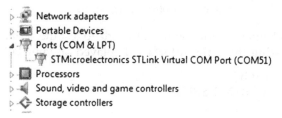

FIG. 8.20 Finding the serial port number.

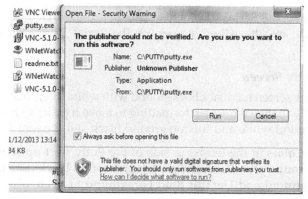

FIG. 8.21 Activate Putty.

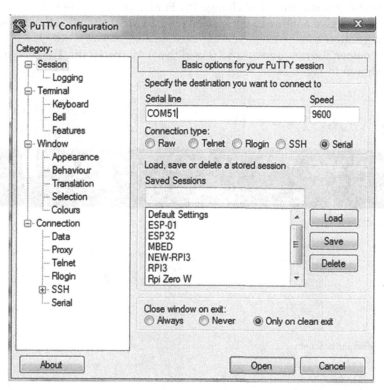

FIG. 8.22 Select Serial line and speed.

- Start your Mbed and enter the following short code to test the serial interface. You should see the message **Testing the serial interface** displayed on your PC screen:

```
#include "mbed.h"
Serial MyPC(USBTX, USBRX);
int main()
{
        MyPC.printf("Testing the serial interface")
}
```

In the above example, the **Serial** statement creates a virtual serial COM port called **MyPC**. **USBTX** and **USBRX** are the Mbed names of the serial port on the Nucleo-F411RE development board.

Configuring the Putty Screen

By default, the Putty screen has black background with white foreground letters as shown in Fig. 8.23. We can easily change the screen formatting to make it easier to read. For example, we can make the background white and the foreground black and the font size to 12 as follows:

- Click **Window ->Colors** at the left-hand side of the Putty start-up screen.
- Set **Default Foreground** and **Default Bold Foreground** to black (Red:0, Green:0, Blue:0)
- Set **Default Background** and **Default Bold Foreground** to white (Red:255, Green:255, Blue:255)
- Set **Cursor Text** and **Cursor Color** to black (Red:0, Green:0, Blue:0)
- Click **Window ->Appearance** and change the **Font size** to 12-point
- Click **Session** and give a name to the session (e.g., STM32) and click **Save** to save the new screen configuration
- Click **Open** to open the new screen and reload the above program. The screen should now display as shown in Fig. 8.24

FIG. 8.23   Default Putty screen.

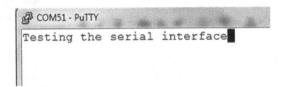

FIG. 8.24   Modified Putty screen.

TABLE 8.2   Serial Communication Functions

Function	Description
pc.printf()	Write a formatted string
pc.putc()	Write a character
pc.getc()	Read a character
pc.scanf()	Read a formatted string
pc.readable()	Determine if there is something to read
pc.writeable()	Determine if it is possible to write
pc.baud()	Set the speed (baud rate) of communication
pc.format()	Set other communication parameters

8.8.2 Mbed Serial Port Functions and Display Control Characters

Mbed supports a number of functions and display control characters that can be useful in PC-based serial display programs. Table 8.2 presents the functions that can be used in serial communication.

printf is probably one of the most commonly used functions as it is used to display text or numeric data on the PC screen. **putc** and **getc** are usually used in pairs to write and read characters from the keyboard, respectively. For example, we can use the **getc** and **putc** function combinations as follows to read and echo characters on the PC screen:

MyPC.putc(MyPC.getc());

scanf is another useful function that can be used to read formatted strings from the keyboard. Functions **readable** and **writeable** determine if it is possible to read or write, respectively. **baud** is used to set the communication speed (default 9600). Finally, **format** sets the other communication parameters such as the data width, parity, etc.

In addition to the serial communication function, we can use the following cursor control characters when sending data to the screen:

\n	Generate newline (at the same column position)
\r	Generate carriage return (beginning of the current line)
\t	Generate tab
\b	Generate backspace

In some applications, we may want to position the cursor at desired coordinates of the screen. Also, we may want to clear the screen or home the cursor position. The following escape codes can be used in such applications (these are also known as VT100 terminal cursor control codes):

esc[H	Home the cursor (top-left position)
esc[2J	Clear the screen
esc[r;cH	Move cursor to row r, column c

Notice that **esc** is the ASCII escape character, having hexadecimal value 0x1B. We can create the following functions to help use the cursor control characters (assuming that the Com port is named MyPC):

```
//
// Clear the screen
//
void clrscr()
{
      char clrscr[] = {0x1B, '[', '2' , 'J',0};
      MyPC.printf(clrscr);
}

//
// Home the cursor
//
void homescr()
{
      char homescr[] = {0x1B, '[' , 'H' , 0};
      MyPC.printf(homescr);
}

//
// Goto specified line and column
//
void gotoscr(int line, int column)
{
      char scr[] = {0x1B, '[', 0x00, ';' ,0x00, 'H', 0};
      scr[2] = line;
      scr[4] = column;
      MyPC.printf(scr);
}
```

8.9 PROJECT 6—VOLUME OF A CYLINDER

8.9.1 Description

In this project, the volume of a cylinder is calculated and displayed on the PC screen. The user enters the radius and height of the cylinder.

8.9.2 Aim

The aim of this project is to show how the various serial communication functions and control codes can be used in a program.

8.9.3 The PDL

The PDL of the program is very simple and is given in Fig. 8.25.

8.9.4 Program Listing

Fig. 8.26 shows the program listing (program: **Cylinder**). At the beginning of the program, variables **radius, height,** and **volume** are defined as float and **Pi** has been initialized to 3.14159. The program then clears the screen, homes the cursor, and displays the heading **Volume of a Cylinder**. The radius and height of the cylinder are then read from the keyboard and the volume is calculated as **Volume = $\pi r^2 h$** and is displayed on the screen. Notice that carriage-return and line-feed characters are sent to the screen before accepting an input from the keyboard so that the messages are displayed on new lines. Also, the user inputs are not echoed on the screen. This is because the local echo is disabled by default on Putty. We can enable the local echo by clicking **Terminal** and then click **Force on** in **Local echo:** as shown in Fig. 8.27.

8.9.5 Suggestions for Additional Work

Modify the program in Fig. 8.26 so that the program displays both the area and the volume of a cylinder.

```
BEGIN
        Clear the screen and home the cursor
        Enter the radius
        Enter the height
        Calculate the volume
        Display the volume
END
```

FIG. 8.25   Program PDL.

```
 COM51 - PuTTY
Volume of a Cylinder
====================
Enter the radius: 10.000000
Enter the height: 5.000000
Volume = 1570.794922
```

```c
/****************************************************************

                  VOLUME OF A CYLINDER
                  ====================

This program calculates and displays the volume of a cylinder.
The user enters the radius and height as floating point values

Author: Dogan Ibrahim
Date  : August 2018
File  : Cylinder
****************************************************************/
#include "mbed.h"
Serial MyPC(USBTX, USBRX);

//
// Clear the screen
//
void clrscr()
{
    char clrscr[] = {0x1B, '[', '2' , 'J',0};
    MyPC.printf(clrscr);
}


//
// Home the cursor
//
void homescr()
{
    char homescr[] = {0x1B, '[' , 'H' , 0};
    MyPC.printf(homescr);
}


int main()
{
    float radius, height, volume;
    float pi = 3.14159;

    clrscr();                                    // Clear the screen
    homescr();                                   // Hoem teh cursor
    MyPC.printf("Volume of a Cylinder");         // Display heading
    MyPC.printf("\n\r====================");
    MyPC.printf("");
    MyPC.printf("\n\rEnter the radius: ");
    MyPC.scanf("%f", &radius);                   // Read the radius
    MyPC.printf("\n\rEnter the height: ");
    MyPC.scanf("%f", &height);                   // Read height
    volume = pi * radius * radius * height;      // Calculate volume
    MyPC.printf("\n\rVolume = %f\n\r", volume);  // Display volume
}
```

FIG. 8.26 Program listing.

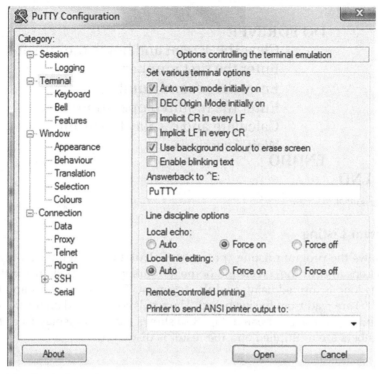

FIG. 8.27 Enabling local echo on Putty.

8.10 PROJECT 7—CALCULATOR

8.10.1 Description

This is a simple calculator program that can perform the basic operations of addition, subtraction, multiplication, and division. The user enters two numbers and the required operation. The result is displayed on the PC screen. The process is repeated after 5 s delay.

8.10.2 Aim

The aim of this project is to show how the various serial communication functions and control codes can be used in a program.

8.10.3 The PDL

The PDL of the program is very simple and is given in Fig. 8.28.

BEGIN
   DO FOREVER
      Clear the screen and home the cursor
      Enter the first number
      Enter the second number
      Enter the required operation
      Calculate and display the result
      Wait 5 seconds
   ENDDO
END

FIG. 8.28   Program PDL.

8.10.4 Program Listing

Fig. 8.29 shows the program listing (program: **Calculator**). At the beginning of the program, the functions to clear the screen and home the cursor are defined. Inside the main program an infinite loop is formed using a while statement. The two numbers and the required operation (+ - * /) are read from the keyboard. The result is displayed on the PC screen for 5 s and then the above process is repeated. Fig. 8.30 shows a typical display from the program where two numbers are multiplied and the result is displayed.

8.11 PROJECT 8—LEARN YOUR TIMES TABLES

8.11.1 Description

This program makes learning the times tables fun. The program asks for a number and then displays the times table for that number from 1 to 12. For example, if the entered number is 5, the program will display the following:

TIMES TABLE FOR 5
==================
$5 \times 1 = 5$
$5 \times 2 = 10$
$5 \times 3 = 15$
$5 \times 4 = 20$
$5 \times 5 = 25$
$5 \times 6 = 30$
$5 \times 7 = 35$
$5 \times 8 = 40$
$5 \times 9 = 45$
$5 \times 10 = 50$
$5 \times 11 = 55$
$5 \times 12 = 60$

```
/******************************************************************

                      CALCULATOR
                      ==========

This is a calculator program. The user enters two numbers and the
requested operation. The result is displayed on the PC screen for
5 seconds and then the process is repeated

Author: Dogan Ibrahim
Date  : August 2018
File  : Calculator
******************************************************************/
#include "mbed.h"
Serial MyPC(USBTX, USBRX);

//
// Clear the screen
//
void clrscr()
{
    char clrscr[] = {0x1B, '[', '2' , 'J',0};
    MyPC.printf(clrscr);
}


//
// Home the cursor
//
void homescr()
{
    char homescr[] = {0x1B, '[' , 'H' , 0};
    MyPC.printf(homescr);
}


int main()
{
    float first, second, result;
    char operation;

    while(1)
    {
        clrscr();                                   // Clear the screen
        homescr();                                  // Home the cursor
        MyPC.printf("\n\rSimple Calculator");       // Display heading
        MyPC.printf("\n\r=================");
        MyPC.printf("");
        MyPC.printf("\n\rEnter First Number: ");
        MyPC.scanf("%f", &first);                   // Read first no
        MyPC.printf("\n\rEnter Second Number: ");
        MyPC.scanf("%f", &second);                  // Read second no
        MyPC.printf("\n\rEnter Operation (+-*/) : ");
        MyPC.scanf("%c", &operation);               // Read operation
```

FIG. 8.29   Program listing.

(Continued)

```
switch (operation)
{
    case '+':                                    // If addition
        result = first + second;
        break;
    case '-':                                    // If subtraction
        result = first - second;
        break;
    case '*':                                    // If multiplication
        result = first * second;
        break;
    case '/':                                    // If division
        result = first / second;
        break;
}

printf("\n\rResult = %f\n", result);             // Display result
wait(5.0);                                       // Wat 5 seconds
}
}
```

FIG. 8.29, CONT'D

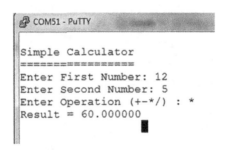

FIG. 8.30 Typical display from the program.

8.11.2 Aim

The aim of this project is to show how the various serial communication functions and control codes can be used in a program.

8.11.3 The PDL

The PDL of the program is very simple and is given in Fig. 8.31.

8.11.4 Program Listing

Fig. 8.32 shows the program listing (program: **Timestable**). At the beginning of the program, the functions to clear the screen and home the cursor are defined. The program then

```
BEGIN
      DO FOREVER
            Clear screen and home cursor
            Enter a number
            Display heading
            DO 12 times
                  Display the times table for the entered number
            ENDDO
      ENDDO
END
```

FIG. 8.31   Program PDL.

asks a number to be entered. The times table for this number is calculated from 1 to 12 and is displayed on the screen. The program clears the screen, homes the cursor, and the process is repeated after 5 s asking for another number to be entered.

A typical display from the program is shown in Fig. 8.33.

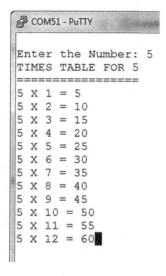

FIG. 8.32   Program listing.

(Continued)

```
/******************************************************************
                        TIMES TABLE
                        ============

This program displays the times table for a given number. The user
enters a number and the program calculates and displays the times
table for this number from 1 to 12. The table is displayed for 5s.
After this time the process is repeated.

Author: Dogan Ibrahim
Date   : August 2018
File   : Timestable
******************************************************************/
#include "mbed.h"
Serial MyPC(USBTX, USBRX);

//
// Clear the screen
//
void clrscr()
{
    char clrscr[] = {0x1B, '[', '2' , 'J',0};
    MyPC.printf(clrscr);
}


//
// Home the cursor
//
void homescr()
{
    char homescr[] = {0x1B, '[' , 'H' , 0};
    MyPC.printf(homescr);
}


int main()
{
   int Number;

   while(1)
   {
       clrscr();                                      // Clear the screen
       homescr();                                     // Home the cursor
       MyPC.printf("\n\rEnter the Number: ");
       MyPC.scanf("%d", &Number);                     // Read the number
       MyPC.printf("");
       MyPC.printf("\n\rTIMES TABLE FOR %d", Number);  // Display heading
       MyPC.printf("\n\r=================");
       MyPC.printf("");
//
// Display the times table for the entered Number
//
       for(int k = 1; k <=12; k++)
       {
           MyPC.printf("\n\r%d X %d = %d", Number, k, k*Number);
       }

       wait(5.0);                                     // Wat 5 seconds
   }
}
```

FIG. 8.32, CONT'D

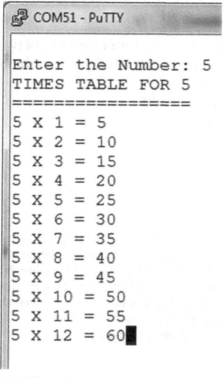

FIG. 8.33   Typical display from the program.

8.12 PROJECT 9—IMPROVING YOUR MULTIPLICATION SKILLS

8.12.1 Description

This program helps the young learners to improve their multiplication skills. The program displays the multiplication of two integer numbers without the result. The numbers are chosen randomly between 1 and 100. The user is given 20 s to calculate the result. After this time, the result is displayed on the screen so that the user can check with his/her own result. The process is repeated after 5 s.

8.12.2 Aim

The aim of this project is to show how the various serial communication functions and control codes can be used in a program.

8.12.3 The PDL

The PDL of the program is shown in Fig. 8.34.

BEGIN
> **DO FOREVER**
>> Generate two random numbers between 1 and 100
>> Display the product of these numbers without the result
>> Wait 20 seconds
>> Display the result
>> Wait 5 seconds
> **ENDDO**

END

FIG. 8.34   Program PDL.

8.12.4 Program Listing

Fig. 8.35 shows the program listing (program: **Multiply**). At the beginning of the program, the functions to clear the screen and home the cursor are defined. The program then generates two random numbers between 1 and 100 and stores in variables Number1 and Number2. The multiplication of these numbers is then shown without the result to give the user to calculate the result. The result is displayed after 20s so that the user can check with his/her own calculation. The process is repeated after 5s delay.

Fig. 8.36 shows the output from the program.

8.13 PROJECT 10—LOOP EXECUTION TIMES

8.13.1 Description

We sometimes need to know the execution times of loops in our programs. In this project, a number of **for** loops are created with different iteration numbers and the loop execution times are calculated and displayed on the PC screen.

8.13.2 Aim

The aim of this project is to show how the Timer function can be used to calculate the execution times of loops in a program.

8.13.3 The PDL

The PDL of the program is shown in Fig. 8.37.

Using a Timer

A **Timer** can be created, started, stopped, and its current value read. It is also permissible to create and use more than one **Timer** in a program. Timers can be useful to measure the elapsed time in a program.

```
/*********************************************************************
                       MULTIPLICATION
                       ==============

This program generates two random numbers between 1 and 100. The
product of these numbers is shown but the result is not shown, and
the user is given 20 seconds to calculate the result. After this
time the result is shown so that the user can check with his/her
own result.

Author: Dogan Ibrahim
Date   : August 2018
File   : Multiply
*********************************************************************/
#include "mbed.h"
Serial MyPC(USBTX, USBRX);

//
// Clear the screen
//
void clrscr()
{
    char clrscr[] = {0x1B, '[', '2' , 'J',0};
    MyPC.printf(clrscr);
}


//
// Home the cursor
//
void homescr()
{
    char homescr[] = {0x1B, '[' , 'H' , 0};
    MyPC.printf(homescr);
}


int main()
{
   int Number1, Number2;

   while(1)
   {
       clrscr();                                            // Clear the screen
       homescr();                                           // Home the cursor
       Number1 = rand() % 100 + 1;                          // First number
       Number2 = rand() % 100 + 1;                          // Second number
       MyPC.printf("\n\r%d X %d = ", Number1, Number2);     // Display the product
       wait(20.0);                                          // Wait 20 seconds
       MyPC.printf("%d\n", Number1 * Number2);              // Display result
       wait(5.0);                                           // Wait 5 seconds
   }
}
```

FIG. 8.35   Program listing.

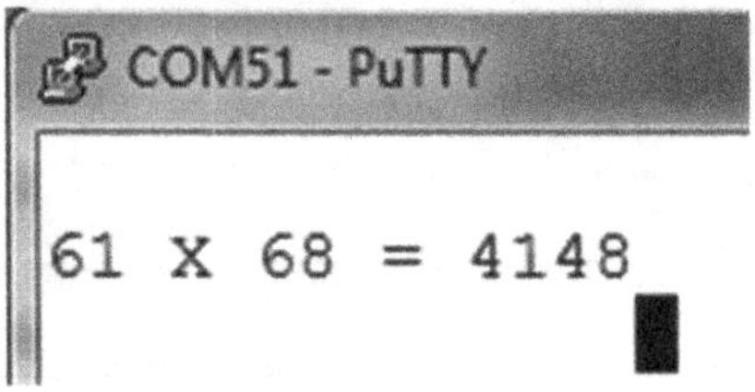

FIG. 8.36 Typical display from the program.

BEGIN
    Display heading
    J = 10,000
    DO 10 times
        Start Timer
            DO J times
            ENDDO
        Stop Timer
        Display J and Elapsed time
        J = 2 * J
    ENDDO
END

FIG. 8.37 Program PDL.

TABLE 8.3 Timer Functions

Timer Function	Description
stop	Stop the timer
start	Start the timer
reset	Reset timer to 0
read	Read the elapsed time in seconds
read_ms	Read the elapsed time in milliseconds
read_us	Read the elapsed time in microseconds

Table 8.3 presents the various **Timer** functions available with Mbed. The following steps show how to measure the elapsed time in part of a program:

- Define a **Timer** variable, for example, **Timer tim**
- Start the **Timer,** for example, **tim.start()**
- Program code whose duration will be measured
- Stop the **Timer,** for example, **tim.stop()**
- Read the elapsed time in seconds, for example, **tim.read()**
- Reset the **Timer to 0,** for example, **tim.reset()**

The elapsed time can be read in seconds, milliseconds, or in microseconds. It is important that the **Timer** should be reset to 0 before it is used successively.

8.13.4 Program Listing

Fig. 8.38 shows the program listing (program: **Loops**). At the beginning of the program, a heading is displayed. Then, two **for** loops are formed. The outer loop counts 10 times. The inner loop is simply a **for** loop with nothing in its body. The **Timer** is started just before entering the loop and it is stopped after exiting from the loop. This inner loop iterates j times where j is initially set to 10,000 and is multiplied every time we enter the inner loop. The time spent in the loop is extracted using function **read_ms** which returns the elapsed time in milliseconds. The loop count (j) and the time spent inside the loop are displayed at every iteration

```
/***********************************************************************

                              LOOP TIMING
                              ===========

This program measures the loop timing of for loops. The iteration
count is changed from 1000 to 512,000 in multiples of 2 and the
time taken to execute the loop is calculated and displayed on the
PC screen in milliseconds.

Author: Dogan Ibrahim
Date  : August 2018
File  : Loops
***********************************************************************/
#include "mbed.h"
Serial MyPC(USBTX, USBRX);
Timer tim;

int main()
{
    unsigned int j;
    float Duration;

    j = 10000;                                               // Starting value

    MyPC.printf("\n\rIteration no     Time(ms)");            // Heading
    MyPC.printf("\n\r============================");

    for(int n = 1; n <= 10; n++)                             // For loop
    {
        tim.reset();                                         // Reset Timer
        tim.start();                                         // Start Timer
        for(int k = 0; k < j; k++)                           // Loop
        {
        }
        tim.stop();                                          // Stop Timer
        Duration = tim.read_ms();                            // Duration (ms)
        MyPC.printf("\n\r%d          %f", j, Duration);      // Display
        j = 2 * j;                                           // New j
    }

    printf("\n\r");                                          // New line
}
```

FIG. 8.38 Program listing.

```
Iteration no       Time(ms)
=============================
10000              0.000000
20000              1.000000
40000              2.000000
80000              4.000000
160000             8.000000
320000            16.000000
640000            32.000000
1280000           64.000000
2560000          128.000000
5120000          256.000000
```

FIG. 8.39   Output from the program.

of the outer loop. It is interesting to note in Fig. 8.38 that the **for** loop takes 8 ms when the iteration count is 160,000.

Fig. 8.39 shows the output from the program.

8.13.5 Suggestions for Additional Work

Modify the program in Fig. 8.38 so that the program displays the elapsed time when the inner loop multiplies two integer numbers.

8.14 PROJECT 11—REACTION TIMER

8.14.1 Description

This is a reaction timer project. The User LED on the Nucleo-F411RE board is lit randomly and the user is expected to press the User button as soon as he/she sees the light coming ON. The time between seeing the light and pressing the button is displayed in milliseconds on the PC screen.

8.14.2 Aim

The aim of this project is to show how the Timer function can be used in a project.

8.14.3 The PDL

The PDL of the program is shown in Fig. 8.40.

8.14.4 Program Listing

Fig. 8.41 shows the program listing (program: **Reaction**). At the beginning of the program, LED1 and BUTTON1 are defined as digital output and digital input, respectively. Inside the main program the following heading is displayed:

Reaction timing
Press the button as soon as you see the light...

BEGIN
    Configure LED1 as output and BUTTON1 as input
    Display heading
    DO FOREVER
        Reset Timer
        Generate a random number 1 to 10 into r
        Wait randomly r seconds
        Start the Timer
        Flash the LED for 50ms
        Wait until button is pressed
        Stop the Timer
        Calculate the elapsed time
        Display the elapsed time
        Wait 3 seconds
    ENDDO
END

FIG. 8.40   Program PDL.

An infinite loop is formed using a **while** statement. Inside this loop, a random number is generated between 1 and 10 and this is used as an argument in the wait statement. Therefore, the program waits randomly between 1 and 10s. The LED is then flashed for 50ms and the program starts the **Timer** and waits until the button is pressed. When the button is pressed, the **Timer** is stopped and the elapsed time is read in milliseconds and stored in variable **Duration**. In all, 50ms is subtracted from the elapsed time since the duration between the ON and OFF times is 50ms. The **Duration** is then displayed on the PC screen. The program repeats after 3s of delay.

Fig. 8.42 shows a typical output from the program.

8.15 PROJECT 12—EVENT COUNTER

8.15.1 Description

This is an event counter project. The User button is used to simulate the occurrence of external events such that every time the button is pressed a count is incremented by one and the total count is displayed on the PC screen continuously.

8.15.2 Aim

The aim of this project is to show how the various serial communication functions can be used.

8.15.3 The PDL

Fig. 8.43 shows the program PDL.

```
/*******************************************************************

                        REACTION TIMING
                        ===============

This program measures the user reaction time. The User LED is lit randomly
and the program waits until the User button is pressed. The elapsed time
between the LED going ON and the button pressed is displayed on the PC
screen in milliseconds. This process is repeated after 3 seconds.

Author: Dogan Ibrahim
Date   : August 2018
File   : Reaction
*******************************************************************/
#include "mbed.h"

Serial MyPC(USBTX, USBRX);
Timer tim;
DigitalOut MyLED(LED1);                              // LED1 is output
DigitalIn button(BUTTON1);                           // BUTTON1 is input

int main()
{
   int r;
   float Duration;


   MyPC.printf("\n\rReaction Timing");               // Display heading
   MyPC.printf("\n\rPress the button as soon as you see the light...");

   while(1)
   {
      tim.reset();                                   // Reset Timer to 0
      r = rand() % 10 + 1;                           // Between 1 and 10
      wait(r);                                       // Random wait
      tim.start();                                   // Start Timer
      MyLED = 1;                                     // LED ON
      wait(0.05);                                    // 50ms delay
      MyLED = 0;                                     // LED OFF
      while(button == 1);                            // Wait for button
      tim.stop();                                    // Stop Timer
      Duration = tim.read_ms() - 50.0;               // Read Duration
      MyPC.printf("\n\rReaction Time (ms) = %f", Duration);
      wait(3.0);
   }
}
```

FIG. 8.41 Program listing.

```
Reaction Timing
Press the button as soon as you see the light...
Reaction Time (ms) = 605.000000
Reaction Time (ms) = 221.000000
Reaction Time (ms) = 142.000000
```

FIG. 8.42 Typical output from the program.

```
BEGIN
        Configure BUTTON1 as digital input
        Clear the screen and home the cursor
        Display heading
        Count = 0
        DO FOREVER
                IF button is pressed THEN
                        Increment Count
                ENDIF
                Go to line 4, column 0
                Display Count =
                Display value of Count
        ENDDO
END
```

FIG. 8.43   Program PDL.

8.15.4 Program Listing

The program listing (program: **SerialEvents**) is shown in Fig. 8.44. At the beginning of the program, **BUTTON1** is configured as digital input, the screen is cleared, cursor is homed, and a heading is displayed. The remainder of the program is executed in an endless loop. Inside this loop, the program waits until the button is pressed. When the button is pressed, the value of **Count** is incremented by 1. The cursor is positioned at line 4, column 0 of the screen and **Count =** is displayed starting from this position. The value of **Count** is then displayed as shown in Fig. 8.45.

8.16 PROJECT 13—HI-LO GAME

8.16.1 Description

This is the popular Hi-Lo game which is played as follows: the computer generates a secret random number. The user guesses this number by entering a guess number. If the guessed number is higher than the secret number, then the message **You are HIGH** is displayed. If on the other hand the guessed number is lower than the secret number, then the message **You are LOW** is displayed. This way the user finally guesses the secret number and the message **Success** is displayed together with the number of attempts made to guess the secret number. In this project, the generated secret number is between 1 and 100.

8.16.2 Aim

The aim of this project is to show how the various serial communication functions can be used in a program.

8.16.3 The PDL

Fig. 8.46 shows the program PDL.

```
/******************************************************************
                        EVENT COUNTER
                        ============

In this program teh user button is used to simulate the occurence
of external events. Every time the button is pressed a count is
incremented by one. The total count is displayed on the PC screen.

Author: Dogan Ibrahim
Date  : August 2018
File  : SerialEvents
******************************************************************/
#include "mbed.h"
Serial MyPC(USBTX, USBRX);
DigitalIn button(BUTTON1);

//
// Clear the screen
//
void clrscr()
{
    char clrscr[] = {0x1B, '[', '2' , 'J',0};
    MyPC.printf(clrscr);
}


//
// Home the cursor
//
void homescr()
{
    char homescr[] = {0x1B, '[' , 'H' , 0};
    MyPC.printf(homescr);
}


//
// Goto specified line and column
//
void gotoscr(int line, int column)
{
    char scr[] = {0x1B, '[', 0x00, ';' ,0x00, 'H', 0};
    scr[2] = line;
    scr[4] = column;
    MyPC.printf(scr);
}


int main()
{
    int Count = 0;                              // Initialzie Count

    clrscr();                                   // Clear the screen
    homescr();                                  // Home the cursor
    MyPC.printf("\n\rEvent Counter");           // Display heading
    MyPC.printf("\n\r==============\n\r");
```

FIG. 8.44   Program listing.

(Continued)

```
    while(1)                                          // Do forever
    {

        while(button == 1);                           // If button is pressed
        Count++;                                       // Increment Count
        while(button == 0);                           // Wait until released

        gotoscr('4', '0');                            // Go to line 4, col 0
        MyPC.printf("Count = ");                      // Display Count =
        MyPC.printf("%d", Count);                     // Display total Count
    }
}
```

FIG. 8.44,   CONT'D

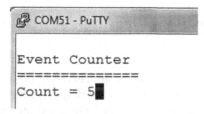

FIG. 8.45   Typical output from the program.

BEGIN
    Initialize USB Serial
    DO FOREVER
        Initialize Attempt count
        Display heading
        Generate a secret random number
        DO FOREVER
            Read user's guess
            Increment Attempt count
            IF guessed number > secret number **THEN**
                Display You are HIGH
            ELSE IF guessed number < secret number **THEN**
                Display You are LOW
            ELSE
                Display Success
                Display Attempt count
            ENDIF
        ENDDO
    ENDDO
END

FIG. 8.46   Program PDL.

8.16.4 Program Listing

The program listing (program: **HiLo**) is shown in Fig. 8.47. The program consists of two nested **while** loops. The outer loop executes indefinitely. The inner loop executes until the secret number has been found (while variable the **FLAG** is 0). Just after the outer loop a heading is displayed and the secret random number is generated between 1 and 100 using the **rand** () function. Inside the inner loop, the user attempts to guess the secret number and receives the messages **You are HIGH, You are LOW**, or **Success** depending on whether or not the guessed number is high, low, or equal to the secret number, respectively. When the user finds the secret number, the number of attempts made to find the number is also displayed after the **Success** message (Fig. 8.48).

8.17 PROJECT 14—SERIAL INTERRUPT

8.17.1 Description

The PC serial communications link can be attached to an ISR so that, for example, when a key is pressed on the keyboard it generates an interrupt and the program jumps to the ISR. In this project, the User LED flashes initially every 100 ms and the rate of flashing is increased by pressing the + key on the keyboard. Pressing the + key generates an interrupt. Inside the ISR the flashing rate is increased by 100 ms to up to 1 s. Pressing the + key again resets the flashing rate back to 100 ms.

8.17.2 Aim

The aim of this project is to show how serial interrupts can be programmed.

8.17.3 The PDL

Fig. 8.49 shows the program PDL. This figure is in two sections: the main program (MAIN) and the ISR.

8.17.4 Program Listing

The program listing (program: **SerISR**) is shown in Fig. 8.50. At the beginning of the program, LED1 is configured as digital output. The main program attaches the serial communication link to a function called **ISR** and it initially flashes the User LED every 100 ms. Inside function ISR, the flashing rate is increased by 100 ms every time the + key is pressed. When the flashing rate is greater than 1 s, it is reset back to 100 ms.

```
/**********************************************************************

                            HiLo Game
                            =========

This is the popular HiLo game. A random secret number is generated by the
computer between 1 and 100 and the user attempts to guess this number. If
the guess is higher than the secret number the message You are HIGH is
displayed. Similarly, if the guess is lower then the message You are LOW is
displayed. The message Success is displayed when the user finds the secret
number. Additionally, the number of attempts is also displayed when the
secret number is found.

Author: Dogan Ibrahim
Date   : August 2018
File   : HiLo
**********************************************************************/
#include "mbed.h"

Serial MyPC(USBTX, USBRX);

int main()
{
    int Flag, SecretNumber, GuessedNumber, Attempts = 0;

    while(1)                                            // Do forever
    {
        Flag = 0;
        MyPC.printf("\n\r");                            // New line
        MyPC.printf("\n\rStart of New Game");           // Display heading
        MyPC.printf("\n\r===================");
        SecretNumber = rand() % 100 + 1;                // Between 1-100

        while(Flag == 0)                                // While not found
        {
            MyPC.printf("\n\rEnter Your Guess (1 - 100): ");
            MyPC.scanf("%d", &GuessedNumber);
            Attempts++;                                 // Increment attempts
            if(GuessedNumber > SecretNumber)            // If higher
                    MyPC.printf("\n\rYou are HIGH");
            else if(GuessedNumber < SecretNumber)       // If lower
                    MyPC.printf("\n\rYou are LOW");
            else                                        // Success
            {
                MyPC.printf("\n\rSuccess...");
                MyPC.printf("Attempts = %d", Attempts); // Attempt count
                Flag = 1;                               // Terminate loop
            }
        }
    }
}
```

FIG. 8.47 Program listing.

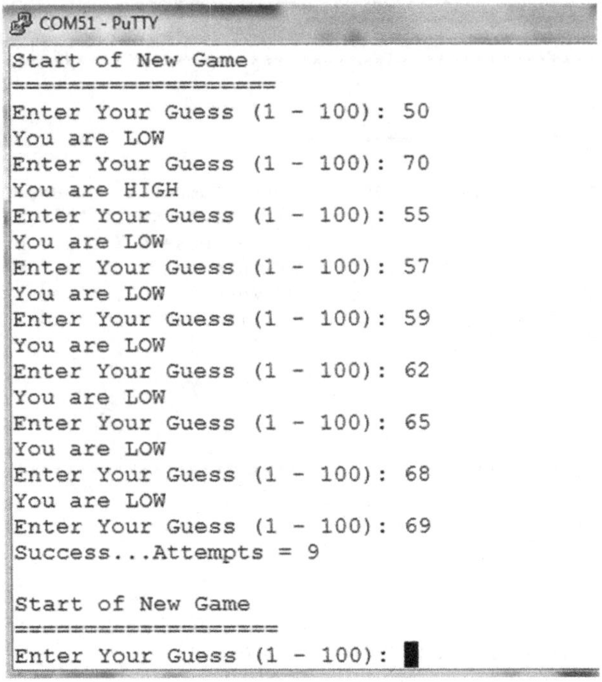

```
COM51 - PuTTY
Start of New Game
====================
Enter Your Guess (1 - 100): 50
You are LOW
Enter Your Guess (1 - 100): 70
You are HIGH
Enter Your Guess (1 - 100): 55
You are LOW
Enter Your Guess (1 - 100): 57
You are LOW
Enter Your Guess (1 - 100): 59
You are LOW
Enter Your Guess (1 - 100): 62
You are LOW
Enter Your Guess (1 - 100): 65
You are LOW
Enter Your Guess (1 - 100): 68
You are LOW
Enter Your Guess (1 - 100): 69
Success...Attempts = 9

Start of New Game
====================
Enter Your Guess (1 - 100): █
```

FIG. 8.48 An example game.

BEGIN/MAIN
    Configure LED1 as output
    Attach serial communication link to function ISR
    dly = 0.1 seconds
    DO FOREVER
        LED ON
        Wait dly seconds
        LED OFF
        Wait dly seconds
    ENDDO
END/MAIN

BEGIN/ISR
    Read a character from the keyboard
    IF the character = '+' **THEN**
        dly = dly + 0.1
        IF dly > 1 second **THEN**
            Dly = 0.1s
        ENDIF
    ENDIF
END/ISR

FIG. 8.49 Program PDL.

```
/*****************************************************************************

                          SERIAL INTERRUPT
                          ================

This program attaches to the serial communication link so that when a
character is typed on the keyboard the program generates an interrupt
and jumps to the interrupt service routine. In this program the User
LED normally flashes every 100 milliseconds. Every time the + key is
pressed on the keyboard the flashing rate increases by 100ms. When
the flashing rate is one second it is reset back to 100ms.

Author: Dogan Ibrahim
Date   : August 2018
File   : SerISR
*****************************************************************************/
#include "mbed.h"

Serial MyPC(USBTX, USBRX);
DigitalOut MyLED(LED1);                         // LED1 is output

#define ON 1
#define OFF 0

double dly;

//
// This is the interrupt service routine. The program jumps here whenever
// a key is pressed on the keyboard. Here, the delay (flashing rate) of the
// LED is increased by 100ms. When the delay is one second it is reset
// back to 100ms
//
void ISR()
{
    char c;
    c == MyPC.scanf("%c",&c);                   // Read the character
    if(c == '+')                                // Is it + ?
    {
        dly = dly + 0.1;                        // Increment dly by 100ms
        if(dly > 1.0)dly = 0.1;                 // If one sec, reset
    }
}

int main()
{
    MyPC.attach(&ISR, Serial::RxIrq);           // Attach to serial link
    dly = 0.1;                                  // Initial value of dly

    while(1)                                    // Do forever
    {
        MyLED = ON;                             // LED ON
        wait(dly);                              // Wait dly
        MyLED = OFF;                            // LED OFF
        wait(dly);                              // Wait dly
    }
}
```

FIG. 8.50   Program listing.

8.18 PROJECT 15—EURO MILLIONAIRE LOTTERY NUMBERS

8.18.1 Description

This is a Euro Millionaire Lottery Numbers project. In the Euro Millions Lottery, there are two sections: the **Lottery Numbers** section and the **Lucky Stars** section where the user is required to select numbers from each section. In the **Lottery Numbers** section, there are numbers from 1 to 50 and the user is required to select only 5 numbers. In the **Lucky Stars** section, there are numbers from 1 to 12 and the user is required to select only 2 numbers. Therefore, 7 numbers have to be selected in total.

In this project, five unique random **Lottery Numbers** are generated between 1 and 50 inclusive, and also two unique random **Lucky Stars** numbers are generated between 1 and 12 inclusive. Clicking the User button generates a set of numbers for the game and these numbers are displayed on the PC screen.

8.18.2 Aim

The aim of this project is to show how the various serial communication functions can be used in a program to generate the Euro Millionaire lottery numbers.

8.18.3 The PDL

Fig. 8.51 shows the program PDL.

8.18.4 Program Listing

The program listing (program: **Lottery**) is shown in Fig. 8.52. At the beginning of the program, two arrays named **LotteryNumbers** and **LuckyStarNumbers** are declared to store the

```
BEGIN/MAIN
        DO FOREVER
                Wait until button is pressed
                Clear the screen and home the cursor
                CALL Generate (5) to generate Lottery numbers
                CALL Generate (2) to generate Lucky Star numbers
                Display the generated Lottery numbers
                Display the generated Luck Star numbers
END/MAIN


BEGIN/Generate (N)
        Generate N unique random numbers
END/Generate
```

FIG. 8.51   Program PDL.

```
/***********************************************************************
                    Euro Millionaire Lottery Numbers
                    ================================

This is a Euro Millionaire Lottery Numbers program. In the Euro Millions
Lottery there are two sections: The Lottery Numbers section and the Lucky
Stars section where the user is required to select numbers from each
section. In this project 5 unique random Lottery Numbers are generated
between 1 and 50 inclusive, and also 2 unique random Lucky Stars numbers
are generated between 1 and 12 inclusive. Clicking the User button
generates a set of numbers for the game and these numbers are displayed
on the PC screen.

Author: Dogan Ibrahim
Date  : August 2018
File  : Lottery
***********************************************************************/
#include "mbed.h"

Serial MyPC(USBTX, USBRX);
DigitalIn button(BUTTON1);                        // BUTTON1 is input

int LotteryNumbers[6],LuckyStarNumbers[3];

//
// Clear the screen
//
void clrscr()
{
    char clrscr[] = {0x1B, '[', '2' , 'J',0};
    MyPC.printf(clrscr);
}

//
// Home the cursor
//
void homescr()
{
    char homescr[] = {0x1B, '[' , 'H' , 0};
    MyPC.printf(homescr);
}

//
// This function generates the Lottery numbers and the Lucky Star
// numbers. The generated numbers are checked to make sure that
// they are unique (i.e. they ar not same as the ones generated in
// this set), otherwise new number is generated
//
void Generate(int Cnt, int Max_No, int LNumbers [])
{
    int i, n, flag, j;

    for(j = 1; j <= Cnt; j++)
```

FIG. 8.52   Program listing.

(Continued)

```
        {
            flag = 1;
            while(flag == 1)
            {
                n = rand() % Max_No + 1;              // Generate a random no
                flag = 0;
                for(i = 1; i <= Cnt; i++)             // Check if repeat
                {
                    if(n == LNumbers[i]) flag = 1;    // If repeat
                }
            }
            LNumbers[j] = n;                           // Save number
        }
    }



int main()
{
    int i, LotteryNoCount, LuckyStarNoCount;
    int MaxLotteryNo, MaxLuckyStarNo;

    while(1)
    {
        while(button == 1);                           // Wait until pressed
        while(button == 0);                           // Wait until released

        clrscr();                                     // Clear screen
        homescr();                                    // Home cursor
//
// Initialize the Lottery and Lucky Star number counts
//
        LotteryNoCount = 5;
        LuckyStarNoCount = 2;
        MaxLotteryNo = 50;
        MaxLuckyStarNo = 12;

//
// Clear arrays LotteryNumbers and LuckyStarNumbers
//
        for(i = 1; i <= LotteryNoCount; i++)
            LotteryNumbers[i] = 0;

        LuckyStarNumbers[1] = 0;
        LuckyStarNumbers[2] = 0;

//
// Generate Lottery Numbers and Lucky Star Numbers
//
        Generate(LotteryNoCount, MaxLotteryNo, LotteryNumbers);
        Generate(LuckyStarNoCount, MaxLuckyStarNo, LuckyStarNumbers);
//
// Display the Lottery Numbers and Luck Star Numbers
```

FIG. 8.52, CONT'D

(Continued)

```
 //
          MyPC.printf("\n\rYour Lottery Numbers are    :   ");        // Heading

               for(i = 1; i <= LotteryNoCount; i++)
                    MyPC.printf("%d ", LotteryNumbers[i]);            // Numbers

          MyPC.printf("\n\rYour Lucky Star Numbers are:   ");         // Heading

               for(i = 1; i <= LuckyStarNoCount; i++)
                    MyPC.printf("%d ", LuckyStarNumbers[i]);          // Numbers
     }
 }
```

FIG. 8.52, CONT'D

generated numbers. The main program executes in an endless loop. Inside this loop, the program waits until the User button is pressed. The two arrays **LotteryNumbers** and **LuckyStarNumbers** are then cleared to 0. Function **Generate** is then called to generate the lottery numbers and the lucky star numbers. Finally, two **for** loops are used to display the generated numbers on the PC screen.

Function **Generate** calls the built-in function **rand()** to generate random numbers for both the lottery numbers and the lucky star numbers. The function checks to make sure that the generated numbers are unique. New numbers are generated if any of the existing numbers are repeated. Fig. 8.53 shows a typical output from the program.

8.19 USING THE ANALOG-TO-DIGITAL CONVERTER

The analog-to-digital converter (ADC) is an important module of a microcontroller. It converts an analog input voltage into a digital number so that it can be processed by the microcontroller or any other digital processor. ADC can be classified into two types as far as the input voltage polarity is concerned. These are unipolar and bipolar. Unipolar ADC accepts unipolar input voltages in the range 0 to +V, and bipolar ADC accepts bipolar input voltages in the range ±V. Bipolar converters are frequently used in signal processing applications, where the signals by nature are bipolar. Unipolar converters are usually cheaper, and they are used in many control and instrumentation applications.

Fig. 8.54 shows the typical steps involved in reading and converting an analog signal into digital form. The front-end of this figure is also known as signal conditioning.

FIG. 8.53   Typical output from the program.

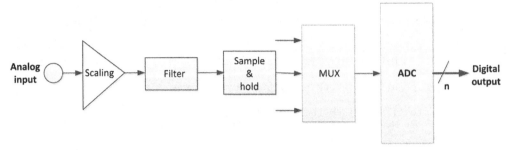

FIG. 8.54   Signal conditioning and ADC process.

Signals received from analog sensors usually need to be processed before being fed to an ADC. This processing usually consists of scaling the input signal to the correct value. Unwanted signal components are then removed by filtering the signal using classical filters (e.g., low-pass filter). The final processing step before feeding the signal to the ADC is to pass it through a sample-and-hold circuit. This is particularly important for fast varying real-time signals whose values may change during the sampling intervals. A sample-and-hold circuit ensures that the signal stays at a constant stable value during the process of conversion. In many analog applications, it is usually required to have more than one ADC. This is normally done using one ADC with a front-end analog multiplexer, where the multiplexer connects only one input signal to the ADC at any one time. The ADC conversion process is as follows:

- Apply the analog signal to one of the ADC inputs
- Start the conversion
- Wait until the conversion is complete (this can be configured to generate an interrupt when the conversion is complete)
- Read the converted digital data

The ADC conversion starts by triggering the converter. Depending on the speed of the ADC, the actual conversion process can take several microseconds. At the end of the conversion, the converter either raises a flag or generates an interrupt to indicate that the conversion is complete and the digital data are available at its outputs.

ADC circuits have reference voltages which determine the step size or the resolution of the conversion process. This is the minimum analog voltage that can be detected at the input. The reference voltage can either be external or internal to the ADC circuit. Most ADCs on microcontrollers are usually 10-bits wide. The ADC on the Nucleo-F411RE development board is a 12-bit converter ($2^{12} = 4096$ quantization levels), having 16 channels and operating with an internal reference voltage of +3.3 V. The ADC resolution is, therefore, given by

$$+3.3\,V/4096 = 0.00080\,V \text{ or } 0.8mV$$

Therefore, for example, if the input voltage is 1 V, the ADC will generate a digital output of $1.0/0.00080 = 1250$ decimal, which is equivalent to "0100 1110 0010" in 12-bit binary. Similarly, if the input voltage is 3.0 V, the converter will generate $3.0/0.00080 = 3750$ decimal, which is equivalent to "1110 1010 0110" in binary.

The 16 ADC channels on the Nucleo-F411RE board are configured as follows:

ADC Channel	GPIO Pin
ADC1/0	PA_0
ADC1/1	PA_1
ADC1/2	PA_2
ADC1/3	PA_3
ADC1/4	PA_4
ADC1/5	PA_5
ADC1/6	PA_6
ADC1/7	PA_7
ADC1/8	PB_0
ADC1/9	PB_1
ADC1/10	PC_0
ADC1/11	PC_1
ADC1/12	PC_2
ADC1/13	PC_3
ADC1/14	PC_4
ADC1/15	PC_5

8.20 PROJECT 16—DIGITAL VOLTMETER

8.20.1 Description

This is a digital voltmeter project. The project measures and displays the voltage applied at one of its analog inputs. The measured voltage is displayed in millivolts on the PC screen. The range of the input voltage must be 0 to +3.3 V. Higher voltages can be measured by using resistive voltage divider circuits at the input of the voltmeter.

8.20.2 Aim

The aim of this project is to show how an analog input port of the Nucleo-F411RE development board can be programmed to measure analog voltages.

8.20.3 Block Diagram

The block diagram of the project is shown in Fig. 8.55. Analog voltage is directly applied to analog input port ADC1/0 (GPIO port PA_0, pin 28 of connector CN7) of the Nucleo-F411RE development board.

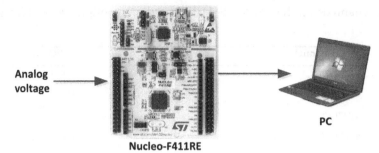

Nucleo-F411RE

FIG. 8.55   Block diagram of the project.

8.20.4 The PDL

Fig. 8.56 shows the program PDL.

8.20.5 Program Listing

Fig. 8.57 shows the program listing (program: **Voltmeter**). At the beginning of the program, analog port ADC1/0 is configured as analog input using the statement **AnalogIn**. The program executes in an endless loop formed using a **while** statement. Inside this loop, the voltage is read using the **read()** statement which returns a floating point data in the range 0.0–1.0. With a 3.3 V ADC reference voltage, 0.0 corresponds to 0 V and 1.0 corresponds to +3.3 V. The actual analog voltage in millivolts is found by multiplying the data read by 3300. This value is then converted into real physical voltage and stored as millivolts in variable **mV**. This value is then displayed on the PC screen. This process is repeated every second. Fig. 8.58 shows a typical output from the program.

```
BEGIN
        Clear the screen and home the cursor
        Display heading
        DO FOREVER
                Read the analog input voltage as digital value
                Convert the reading into voltage in mV
                Display the voltage in mV
                Wait one second
        ENDDO
END
```

FIG. 8.56   Program PDL.

```
/************************************************************************

                            VOLTMETER
                            =========

This is a voltmeter program. Input analog voltage is applied to analog
input ADC1/0 (GPIO pin PA_0) of the Nucleo-F411RE development board.
The program reads and displays the analog voltage in millivolts on the
PC screen

Author: Dogan Ibrahim
Date   : August 2018
File   : Voltmeter
*************************************************************************/
#include "mbed.h"

Serial MyPC(USBTX, USBRX);
AnalogIn ain(PA_0);

// Clear the screen
//
void clrscr()
{
    char clrscr[] = {0x1B, '[', '2' , 'J',0};
    MyPC.printf(clrscr);
}


//
// Home the cursor
//
void homescr()
{
    char homescr[] = {0x1B, '[' , 'H' , 0};
    MyPC.printf(homescr);
}


//
// Goto specified line and column
//
void gotoscr(int line, int column)
{
    char scr[] = {0x1B, '[', 0x00, ';' ,0x00, 'H', 0};
    scr[2] = line;
    scr[4] = column;
    MyPC.printf(scr);
}

int main()
{
    double mV;

    clrscr();                                   // Clear the screen
    homescr();                                  // Home teh cursor
    MyPC.printf("\n\rVOLTMETER");               // Heading
    MyPC.printf("\n\r=========");
```

FIG. 8.57   Program listing.

(Continued)

```
    while(1)                                    // Do forever
    {
        mV = 3300.0f * ain.read();              // Voltage in mV
        gotoscr('4', '0');                      // Goto line 4 col 0
        MyPC.printf("Voltage = ");
        MyPC.printf("%f", mV);                  // Display voltage
        wait(1.0);                              // Wait 1 second
    }
}
```

FIG. 8.57, CONT'D

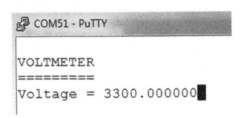

FIG. 8.58   Typical output from the program.

8.21 PROJECT 17—ANALOG TEMPERATURE SENSOR (DIGITAL THERMOMETER)

8.21.1 Description

In this project, the ambient temperature is measured using an analog temperature sensor chip, and is then displayed on the PC screen every second.

8.21.2 Aim

The aim of this project is to show how the ambient temperature can be measured using a temperature sensor chip.

8.21.3 Block Diagram

Fig. 8.59 shows the block diagram of the project. In this project, a TMP36-type analog temperature chip is used. This is a low-voltage precision temperature sensor (see Fig. 8.60) having the following features:

- Voltage input: 2.7V–5.5VDC
- 10mV/°C scale factor
- ±2°C accuracy over temperature
- ±0.5°C linearity
- Operating range: −40°C to +125°C
- Less than 50µA quiescent current

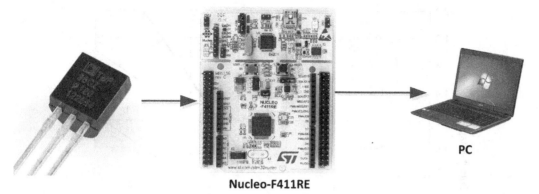

Nucleo-F411RE

FIG. 8.59   Block diagram of the project.

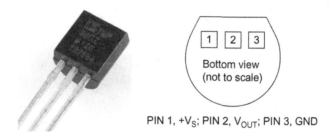

Bottom view
(not to scale)

PIN 1, +V_S; PIN 2, V_{OUT}; PIN 3, GND

FIG. 8.60   TMP36 temperature sensor chip.

The output voltage of TMP36 is linearly proportional to the measured temperature (see Fig. 8.61, curve b) and is given by

$$T = (V_o - 500)/10$$

where T is the measured temperature in degrees Centigrade and V_o is the sensor output voltage in millivolts. For example, if the measured voltage is 800 mV, then the temperature is

$$T = (800 - 500)/10 = 30°C$$

8.21.4 Circuit Diagram

The circuit diagram of the project is shown in Fig. 8.62. The output of the sensor is directly connected to analog input ADC1/0 (GPIO port PA_0, pin 28 of connector CN7) of the Nucleo-F411RE development board. The +Vs power supply input of the sensor chip is connected to +3.3 V (pin 16 of connector CN7) of the development board. Similarly, the ground pin of the sensor chip is connected to the GND pin (pin 8 of connector CN7) of the development board.

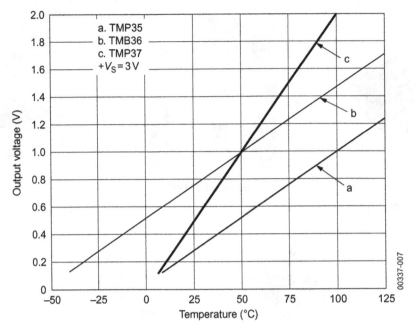

FIG. 8.61   TMP36 characteristics.

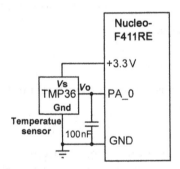

FIG. 8.62   Circuit diagram of the project.

8.21.5 The Construction

The project is constructed on a breadboard as shown in Fig. 8.63. The sensor chip is connected to the development board using jumper wires.

8.21.6 The PDL

Fig. 8.64 shows the program PDL.

8.21.7 Program Listing

The program listing (program: **TMP36**) is shown in Fig. 8.65. At the beginning of the program, analog port ADC1/0 is configured as analog input using the statement **AnalogIn**.

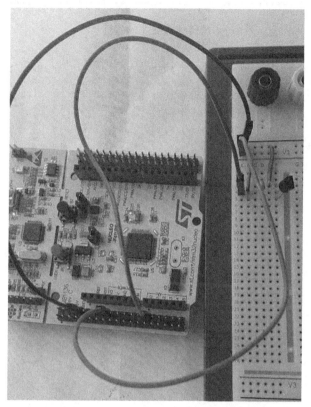

FIG. 8.63   The project on a breadboard.

BEGIN
        Clear the screen and home the cursor
        Display heading
        DO FOREVER
                Read the sensor voltage as digital value
                Convert the reading into voltage in mV
                Convert the reading into degrees Centigrade
                Display the temperature in degrees Centigrade
                Wait one second
        ENDDO
END

FIG. 8.64   Program PDL.

```
/*************************************************************************

                          DIGITAL THERMOMETER
                          ===================

This is a voltmeter program. Input analog voltage is applied to analog
input ADC1/0 (GPIO pin PA_0) of the Nucleo-F411RE development board.
The program reads and displays the analog voltage in millivolts on the
PC screen

Author: Dogan Ibrahim
Date   : August 2018
File   : TMP36
*************************************************************************/
#include "mbed.h"

Serial MyPC(USBTX, USBRX);
AnalogIn ain(PA_0);

// Clear the screen
//
void clrscr()
{
    char clrscr[] = {0x1B, '[', '2' , 'J',0};
    MyPC.printf(clrscr);
}


//
// Home the cursor
//
void homescr()
{
    char homescr[] = {0x1B, '[' , 'H' , 0};
    MyPC.printf(homescr);
}


//
// Goto specified line and column
//
void gotoscr(int line, int column)
{
    char scr[] = {0x1B, '[', 0x00, ';' ,0x00, 'H', 0};
    scr[2] = line;
    scr[4] = column;
    MyPC.printf(scr);
}

int main()
{
    double mV, T;

    clrscr();                                  // Clear the screen
    homescr();                                 // Home teh cursor
    MyPC.printf("\n\rDIGITAL THERMOMETER");    // Heading
    MyPC.printf("\n\r===================");
```

FIG. 8.65 Program listing.

(Continued)

```
    while(1)                              // Do forever
    {
        mV = 3300.0f * ain.read();        // Voltage in mV
        T = (mV - 500.0f) / 10.0f;        // Temperature in C
        gotoscr('4', '0');                // Goto line 4 col 0
        MyPC.printf("Temperature = ");
        MyPC.printf("%5.2f", T);          // Display voltage
        wait(1.0);                        // Wait 1 second
    }
}
```

FIG. 8.65,   CONT'D

The program executes in an endless loop formed using a **while** statement. Inside this loop, the output voltage of the sensor is read using the **read()** statement which returns a floating point data in the range 0.0–1.0. With a 3.3 V ADC reference voltage, 0.0 corresponds to 0 V and 1.0 corresponds to +3.3 V. The actual analog voltage in millivolts is found by multiplying the data read by 3300. This value is then converted into real physical voltage and stored as millivolts in variable **mV**. The voltage reading is then converted into degrees Centigrade temperature by subtracting 500 and dividing by 10. This value is then displayed on the PC screen every second. Fig. 8.66 shows a typical output from the program.

8.22 PROJECT 18—DIGITAL THERMOSTAT

8.22.1 Description

This is a thermostat project. The project measures and displays the ambient temperature every second on the PC screen. The user enters a maximum and minimum temperature values through the keyboard. If the temperature is within the specified limits, then the message **NORMAL TEMPERATURE** is displayed together with the actual temperature reading. If the temperature goes above or below the set points, then the user LED on the development board is lit to indicate an alarm condition. At the same time, the messages **ALARM—LOW TEMPERATURE** or **ALARM—HIGH TEMPERATURE** are displayed on the screen together with the actual temperature readings.

8.22.2 Aim

The aim of this project is to show how the ambient temperature can be measured and how a thermostat can be designed to give alarm conditions if the temperature is outside the specified limits.

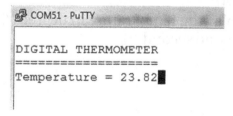

FIG. 8.66   Typical output from the program.

8.22.3 Block Diagram

The block diagram of the project is as in Fig. 8.59.

8.22.4 Circuit Diagram

The circuit diagram of the project is as in Fig. 8.62. The User LED on the development board is used to indicate an alarm condition. No other external components are used in the design.

8.22.5 The PDL

Fig. 8.67 shows the program PDL.

8.22.6 Program Listing

Fig. 8.68 shows the program listing (program: **Thermostat**). At the beginning of the program, the screen is cleared, cursor is homed, and the heading DIGITAL THERMOSTAT is displayed. The program then reads the desired maximum and minimum temperature values and stores them in variables MaxTemp and MinTemp, respectively. The remainder of the program is executed in an endless loop. Inside this loop, the temperature is read every second

```
BEGIN
        Clear the screen and home the cursor
        Display heading
        Read the desired max and min temperatures
        DO FOREVER
                Read the sensor voltage as digital value
                Convert the reading into voltage in mV
                Convert the reading into degrees Centigrade
                IF temperature > desired value THEN
                        Display ALARM-HIGH TEMPERATURE
                ELSE IF temperature < desired value THEN
                        Display ALARM-LOW_TEMPERATURE
                ELSE IF temperatures between desired values THEN
                        Display NORMAL TEMPERATURE
                ENDIF
                Display the temperature in degrees Centigrade
                Wait one second
        ENDDO
END
```

FIG. 8.67   Program PDL.

```
/************************************************************************

                        DIGITAL THERMOSTAT
                        ==================

This is a digital thermostat program. The user enters the desired Max
and Min temperatures from the keyboard. The program measures the ambient
temperature every second. If the temperature is above the Max value then
the message ALARM-HIGH TEMPERATURE is displayed and the User LED is lit.
Also, if the temperature is below the Min value then the message ALARM-
LOW TEMPERATURE is displayed and the LED is lit to indicate the alarm
condition. If the temperature is within the desired limits then the
message MORMAL TEMPERATURE is displayed and the LED is turned OFF.

Author: Dogan Ibrahim
Date   : August 2018
File   : Thermostat
*************************************************************************/
#include "mbed.h"

Serial MyPC(USBTX, USBRX);
AnalogIn ain(PA_0);
DigitalOut MyLED(LED1);

// Clear the screen
//
void clrscr()
{
    char clrscr[] = {0x1B, '[', '2' , 'J',0};
    MyPC.printf(clrscr);
}


//
// Home the cursor
//
void homescr()
{
    char homescr[] = {0x1B, '[' , 'H' , 0};
    MyPC.printf(homescr);
}


//
// Goto specified line and column
//
void gotoscr(int line, int column)
{
    char scr[] = {0x1B, '[', 0x00, ';' ,0x00, 'H', 0};
    scr[2] = line;
    scr[4] = column;
    MyPC.printf(scr);
}

int main()
{
```

FIG. 8.68 Program listing.

(Continued)

```
    double mV, T;
    float MaxTemp, MinTemp;

    clrscr();                                    // Clear the screen
    homescr();                                   // Home teh cursor
    MyPC.printf("\n\rDIGITAL THERMOSTAT");   // Heading
    MyPC.printf("\n\r===================");
//
// Read the desired Max and Min temperatures
//
    MyPC.printf("\n\rEnter Max Temperature : ");
    MyPC.scanf("%f", &MaxTemp);
    MyPC.printf("\n\rEnter Min Temperature : ");
    MyPC.scanf("%f", &MinTemp);
//
// Read and convert the temperature into Degrees C. Set the alarm
// conditions if the temperature measured is low or high
//
    while(1)                                         // Do forever
    {
        mV = 3300.0f * ain.read();                   // In mV
        T = (mV - 500.0f) / 10.0f;                   // In Degrees C
        gotoscr('7', '0');                           // Line 7 col 0

        if(T < MaxTemp && T > MinTemp)               // If normal
        {
            MyPC.printf("NORMAL TEMPERATURE      = ");
            MyLED = 0;                               // LED OFF
        }
        else if(T >= MaxTemp)                        // If above
        {
            MyPC.printf("ALARM-HIGH TEMPERATURE = ");
            MyLED = 1;                               // LED ON
        }
        else if(T <= MinTemp)                        // If below
        {
            MyPC.printf("ALARM-LOW TEMPERATURE   = ");
            MyLED = 1;                               // LED ON
        }
        MyPC.printf("%5.2f", T);                     // Display T

        wait(1.0);                                   // Wait 1 second
    }
}
```

FIG. 8.68, CONT'D

and converted into degree Centigrade. The temperature is compared with the desired values and appropriate messages are displayed on the PC screen. In addition, the current reading of the temperature is displayed as a floating point number in the format **%5.2f**, which displays the data as **nn·nn**. The User LED is turned ON if the temperature is below or above the desired values (Figs. 8.69 and 8.70).

COM51 - PuTTY

DIGITAL THERMOMETER
====================
Enter Max Temperature : 27
Enter Min Temperature : 18

NORMAL TEMPERATURE   = 19.79

FIG. 8.69 Typical output from the program (normal run).

COM51 - PuTTY

DIGITAL THERMOMETER
====================
Enter Max Temperature : 15
Enter Min Temperature : 12

ALARM-HIGH TEMPERATURE = 19.63

FIG. 8.70 Typical output from the program (alarm condition).

8.22.7 Modified Project

The project can be modified by adding a buzzer to one of the GPIO ports so that the buzzer sounds when an alarm condition occurs. There are two types of buzzers: **passive** and **active**. Passive buzzers are driven by audible signals and they can generate sounds at different frequencies. These types of buzzers are normally driven using pulse width modulation (PWM) signals or sine wave signals. Active buzzers on the other hand are driven by logic signals and sending logic 1 activates the buzzer where the buzzer sounds at a fixed audible frequency. In the modified project, an active buzzer is connected to port PC_0 of the development board as shown in Fig. 8.71. The program (program: **Thermostat-2**) configures the buzzer as digital

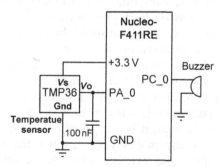

FIG. 8.71 Modified circuit diagram.

output and activates the buzzer when an alarm condition occurs. The following changes are required in the main program:

```
DigitalOut Buzzer(PC_0);
..............................
..............................
if(T < MaxTemp && T > MinTemp)
{
        MyPC.printf("NORMAL TEMPERATURE   = ");
        Buzzer = 0;
}
else if(T >= MaxTemp)
{
        MyPC.printf("ALARM-HIGH TEMPERATURE = ");
        Buzzer = 1;
}
else if(T <= MinTemp)
{
        MyPC.printf("ALARM-LOW TEMPERATURE  = ");
        Buzzer = 1;
}
```

8.22.8 Suggestions for Additional Work

Modify the circuit given in Fig. 8.71 by replacing the buzzer with a relay so that the relay is activated when an alarm condition occurs.

8.23 PROJECT 19—ON/OFF TEMPERATURE CONTROLLER

8.23.1 Description

This is an ON/OFF temperature controller project. The ambient temperature in a room, or the temperature of an oven, or of a liquid can be controlled using two methods: ON/OFF-type control and continuous control (e.g., using PID, Proportional + Integral + Derivative algorithm or a similar algorithm). In both methods, the temperature is measured using a sensor and negative feedback is applied to bring the temperature to the desired set point.

In this project, ON/OFF-type control is used where the desired set-point temperature is compared with the actual measured temperature. A heater is connected to a relay where the relay is turned ON or OFF accordingly so that the desired temperature is achieved and maintained.

In practice, ON/OFF-type controller is the simplest form of feedback control, but it has some drawbacks. First, the actuator (e.g., relay) is turned ON and OFF so many times that this may shorten its life time. Second, it is not possible to achieve precision control with the ON/OFF-type controller. Usually, in this type of control, the desired temperature fluctuates around the desired set point as the relay is turned ON and OFF. But, ON/OFF-type control is still used in many simple temperature control applications where a few degrees change from the set point can be tolerated.

8.23.2 Aim

The aim of this project is to show how an ON/OFF-type temperature controller system can be designed.

8.23.3 Block Diagram

The block diagram of the project is as in Fig. 8.72. The system consists of the Nucleo-F411RE development board, a temperature sensor chip, a relay, and an electric heater.

8.23.4 Circuit Diagram

The circuit diagram of the project is shown in Fig. 8.73. The output of a TMP36-type temperature sensor chip is connected to ADC1/0 (GPIO port PA_0, pin 28 of connector CN7) pin of the Nucleo-F411RE development board. The +Vs power supply input of the sensor chip is connected to +3.3V (pin 16 of connector CN7) of the development board. Similarly, the ground pin of the sensor chip is connected to the GND pin (pin 8 of connector CN7) of the

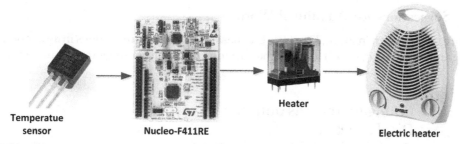

Temperatue sensor   **Nucleo-F411RE**   **Heater**   **Electric heater**

FIG. 8.72   Block diagram of the project.

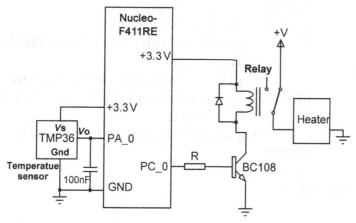

FIG. 8.73   Circuit diagram of the project.

development board. A relay is connected to GPIO pin PC_0 through a bipolar transistor switch. The heater is connected to one of the relay contacts and to the power supply as shown in the figure.

8.23.5 The PDL

Fig. 8.74 shows the program PDL.

8.23.6 Program Listing

Fig. 8.75 shows the program listing (program: **ONOFF**). At the beginning of the program, a heading is displayed and the desired set-point temperature is read from the keyboard and stored in a floating point variable called **SetPoint**. The remainder of the program runs in an endless loop. Inside this loop, the ambient temperature is measured and compared with the set point. The relay is turned ON or OFF depending on whether the set point is below or above the measured value, respectively. Both the desired set point and the measured temperatures are displayed on the screen at line 7, column 0. The above process is repeated every second.

Fig. 8.76 shows a typical run of the program.

8.23.7 Suggestions for Additional Work

Modify the program in Fig. 8.75 to save the measured temperature values in an array. Then, draw a graph (e.g., using Excel) to show the variation of the measured temperature with time.

```
BEGIN
        Configure relay as output
        Display heading
        Read the desired Set-Point temperature
        DO FOREVER
                Read the sensor voltage
                Convert to Degrees centigrade
                IF measured temperature < Set-Point THEN
                        Turn ON relay
                        Display message that the relay is ON
                ELSE IF measured temperature > Set-Point THEN
                        Turn OFF relay
                        Display message that the relay is OFF
                ENDIF
                Wait 1 second
        ENDDO
END
```

FIG. 8.74 Program PDL.

```
/*************************************************************************

                    ON/OFF TEMPERATURE CONTROLLER
                    =============================

This is an ON/OFF temperature controller program. The program reads the
desired set-point temperature from the keyboard. A temperature sensor, a
relay and heater are connected to the development board. If the measured
temperature is above the desired set-point then the heater is turned OFF.
If on the other hand the measured temperature is below the set-point
temperature then the heater is turned ON. The program displays both the
set-point and the measured temperatures on the PC screen.

Author: Dogan Ibrahim
Date  : August 2018
File  : ONOFF
*************************************************************************/
#include "mbed.h"

Serial MyPC(USBTX, USBRX);
AnalogIn ain(PA_0);
DigitalOut Relay(PC_0);

// Clear the screen
//
void clrscr()
{
    char clrscr[] = {0x1B, '[', '2' , 'J',0};
    MyPC.printf(clrscr);
}


//
// Home the cursor
//
void homescr()
{
    char homescr[] = {0x1B, '[' , 'H' , 0};
    MyPC.printf(homescr);
}


//
// Goto specified line and column
//
void gotoscr(int line, int column)
{
    char scr[] = {0x1B, '[', 0x00, ';' ,0x00, 'H', 0};
    scr[2] = line;
    scr[4] = column;
    MyPC.printf(scr);
}


int main()
{
```

FIG. 8.75   Program listing.

(Continued)

```
            double mV, T;
            float SetPoint;

            clrscr();                                              // Clear the screen
            homescr();                                             // Home the cursor
            MyPC.printf("\n\rON/OFF TEMPERATURE CONTROLLER");      // Display heading
    //
    // Read the desired Set-Point temperature
    //
            MyPC.printf("\n\rEnter the Desired Set-Point Temperature : ");
            MyPC.scanf("%f", &SetPoint);

    //
    // Read and convert the temperature into Degrees C. Display the
    // Set-Point and the measured temperatures. If the measured value
    // is greater than the Set-Point then turn OFF the relay. If on the
    // other hand the measured value is lower than the Set-Point then
    // turn ON the relay
    //
            while(1)                                               // Do forever
            {
                mV = 3300.0f * ain.read();                         // In mV
                T = (mV - 500.0f) / 10.0f;                         // In Degrees C
                gotoscr('7', '0');                                 // Line 7 col 0

                MyPC.printf("Set-Point Temperature = %5.2f", SetPoint);
                MyPC.printf("\n\rMeasured Temperature  = %5.2f", T);

                if(T < SetPoint)                                   // If higher
                {
                    Relay = 1;                                     // Relay ON
                    MyPC.printf("\n\rRELAY IS ON ");
                }
                else if(T > SetPoint)
                {
                    Relay = 0;
                    MyPC.printf("\n\rRELAY IS OFF");
                }

                wait(1.0);                                         // Wait 1 second
            }
        }
```

FIG. 8.75, CONT'D

```
 COM51 - PuTTY

ON/OFF TEMPERATURE CONTROLLER
Enter the Desired Set-Point Temperature : 24




Set-Point Temperature = 24.00
Measured Temperature  = 19.71
RELAY IS ON █
```

FIG. 8.76 Typical run of the program.

8.24 PROJECT 20—LIGHT LEVEL METER

8.24.1 Description

This is a light level meter which measures the ambient light level (illumination) in lux and displays the readings on the PC screen. A light-dependent resistor (LDR) is used in this project.

8.24.2 Aim

The aim of this project is to show how a LDR can be used in a project to measure and display the ambient light level in lux.

8.24.3 Block Diagram

In this project, the NORPS-12-type LDR is used (see Fig. 8.77). This device is based on CdS photoconductive cell and is sensitive to visible light in the region of 550 nm. The resistance of the device increases as the light level is reduced. As shown in Fig. 8.78, the resistance is greater than 1 mega-ohm in the dark, and it falls to around 1K in the light. The block diagram of the project is as in Fig. 8.79.

8.24.4 Circuit Diagram

The circuit diagram of the project is shown in Fig. 8.80. A resistive voltage divider circuit is formed by connecting the LDR to the power supply through a 10K resistor. The voltage across the LDR is connected to ADC1/0 (GPIO port PA_0, pin 28 of connector CN7) pin of the Nucleo-F411RE development board. The program measures the voltage across the LDR, calculates the resistance of the LDR, and displays it on the PC screen.

FIG. 8.77   The NORPS-12 LDR.

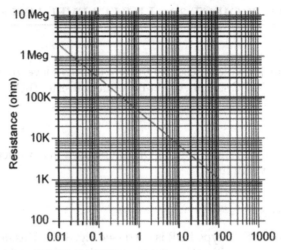

FIG. 8.78   Characteristic of the NORPS-12 LDR (www.lunainc.com).

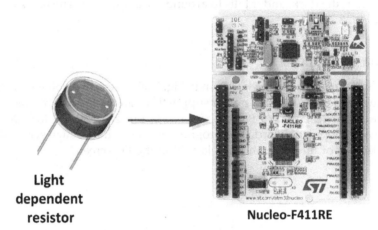

FIG. 8.79   Block diagram of the project.

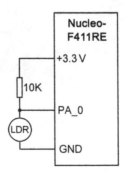

FIG. 8.80   Circuit diagram of the project.

The resistance of the LDR is calculated as:

Assuming that R is the resistance used and the resistance of the LDR is R_L,

$$V = \frac{3300R_L}{R + R_L} \tag{8.1}$$

where V is the voltage measured across the LDR (in mV). The LDR resistance is then found as

$$R_L = \frac{VR}{3300 - V} \tag{8.2}$$

R is 10K, therefore, the equation becomes

$$R_L = \frac{10V}{3300 - V} \tag{8.3}$$

where R_L is the measured LDR resistance in kilo-ohms and V_o is in millivolts.

The relationship between the LDR resistance of the NORPS-12 and the illumination is given approximately by

$$\text{Log}_{10}(L) = 2.17 - 1.28 \times \text{Log}_{10}(R_L) \tag{8.4}$$

where L is the illumination in lux and R_L is as before (i.e., the LDR resistance in kilo-ohms). Therefore, we can first find the resistance R_L using Eq. (8.3). Then, we can calculate the illumination L from Eq. (8.4).

8.24.5 The Construction

The project is constructed on a breadboard as shown in Fig. 8.81.

8.24.6 The PDL

Fig. 8.82 shows the program PDL.

8.24.7 Program Listing

Fig. 8.83 shows the program listing (program: **LDR**). At the beginning of the program, a heading is displayed. The program then reads the voltage across the LDR and stores in variable **mV**. The resistance of the LDR is then calculated and stored in variable **R**. The program finally calculates the light level in lux and displays it at line 4, column 0 of the PC screen. The display is refreshed every second. Fig. 8.84 shows a typical display of the light level. Notice that it will be required to calibrate the readings for accurate measurements using a professional light level meter.

Notice that function **log10** calculates the logarithm to base 10 of the given number. There is no antilogarithm function in Mbed. But since antilogarithm is same as raising 10 to the power of the given number, the **pow()** function is used to calculate the antilogarithm and hence the ambient light level.

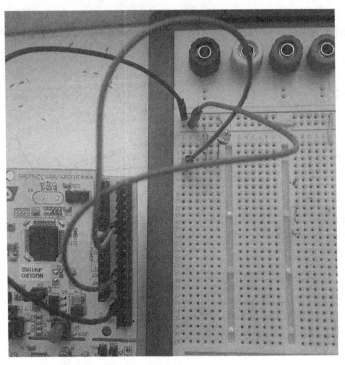

FIG. 8.81   Project constructed on a breadboard.

BEGIN
        Display heading
        DO FOREVER
            Read voltage across LDR
            Calculate LDR resistance
            Calculate the light level
            Display the light level in lux
            Wait 1 second
        ENDDO
    END

FIG. 8.82   Program PDL.

8.25 PROJECT 21—CHANGING LED FLASHING RATE WITH A POTENTIOMETER

8.25.1 Description

In this project, a potentiometer is used to change the flashing rate of the User LED. As the potentiometer arm is rotated, the LED flashing rate changes from 0 (no flashing) to flashing every second.

```
/***********************************************************************

                        LIGHT LEVEL METER
                        =================

This is a light level meter project. A light dependent resistor is
connected to one of the analog inputs of the development board. The
project displays the light level in lux on the PC screen.

Author: Dogan Ibrahim
Date   : August 2018
File   : LDR
***********************************************************************/
#include "mbed.h"

Serial MyPC(USBTX, USBRX);
AnalogIn LDR(PA_0);

// Clear the screen
//
void clrscr()
{
    char clrscr[] = {0x1B, '[', '2' , 'J',0};
    MyPC.printf(clrscr);
}


//
// Home the cursor
//
void homescr()
{
    char homescr[] = {0x1B, '[' , 'H' , 0};
    MyPC.printf(homescr);
}


//
// Goto specified line and column
//
void gotoscr(int line, int column)
{
    char scr[] = {0x1B, '[', 0x00, ';' ,0x00, 'H', 0};
    scr[2] = line;
    scr[4] = column;
    MyPC.printf(scr);
}


int main()
{
    double mV, R, lux;

    clrscr();                                    // Clear the screen
    homescr();                                   // Home the cursor
    MyPC.printf("\n\rLIGHT LEVEL METER");        // Display heading
    MyPC.printf("\n\r=================");
```

FIG. 8.83 Program listing.

(Continued)

```
//
// Read the output voltage of the LDR, then calculate the resistance
// of it and finally calculate and display the light level in lux
//
   while(1)                                          // Do forever
   {
       mV = 3300.0f * LDR.read();                     // In mV
       R = 10.0 * mV / (3300.0 - mV);                 // LDR resistance
       lux = 2.17 - 1.28 * log10(R);                  //
       lux = pow(10.0, lux);                          // Light level (Lux)
       gotoscr('4', '0');                             // Line 4, col 0
       MyPC.printf("Light Level (lux) = %5.2f", lux);

       wait(1.0);                                     // Wait 1 second
   }
}
```

FIG. 8.83, CONT'D

```
COM51 - PuTTY

LIGHT LEVEL METER
==================
Light Level (Lux) = 24.55
```

FIG. 8.84   Typical display of the light level.

8.25.2 Aim

The aim of this project is to show how the flashing rate of the User LED can be changed by using a potentiometer.

8.25.3 Block Diagram

The block diagram of the project is shown in Fig. 8.85.

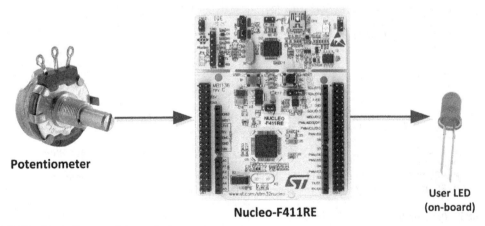

Potentiometer

Nucleo-F411RE

User LED (on-board)

FIG. 8.85   Block diagram of the project.

8.25.4 Circuit Diagram

The arm of a 10K potentiometer is connected to analog input ADC1/0 (GPIO port PA_0, pin 28 of connector CN7) of the Nucleo-F411RE development board as shown in Fig. 8.86. The other two pins of the potentiometer are connected to +3.3 V and to the ground pins, respectively.

8.25.5 The PDL

The PDL of the program is shown in Fig. 8.87.

8.25.6 Program Listing

Fig. 8.88 shows the program listing (program: **POTLED**). The program reads the analog voltage at the arm of the potentiometer. This voltage varies between 0 and +3.3 V. Function **pot.read()** returns the voltage as a floating point number in variable **Dig** between 0 and 1.0 (multiplying this number by 3.3 gives the actual physical analog voltage read). This reading is used as the delay parameter in a **wait** statement so that the delay varies between 0 and 1.0 s as the potentiometer arm is rotated. This delay is used in flashing the LED ON and OFF.

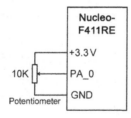

FIG. 8.86   Circuit diagram of the project.

```
BEGIN
        Configure User LED as output
        DO FOREVER
                Read potentiometer arm voltage into variable Dig
                Turn LED ON
                Wait Dig seconds
                Turn LED OFF
                Wait Dig seconds
        ENDO
END
```

FIG. 8.87   PDL of the program.

```
/***********************************************************************

                    FLASHING LED WITH POTENTIOMETER
                    ===============================

In this project the arm of s potentiometer is connected to analog input
PA_0 of teh Nucleo-F411RE development board. The User LED on the board
is flashed where the flashing rate is controlled by the potentiometer as
the potentiometer arm is rotated. The flashing rate is 0 (no flashing)
to flashing every second

Author: Dogan Ibrahim
Date   : August 2018
File   : POTLED
***********************************************************************/
#include "mbed.h"

Serial MyPC(USBTX, USBRX);
AnalogIn pot(PA_0);
DigitalOut MyLED(LED1);

#define ON 1
#define OFF 0

int main()
{
    float Dig;

    while(1)                                // Do forever
    {
        Dig = pot.read();                   // In mV
        MyLED = ON;                         // LED ON
        wait(Dig);                          // Delay
        MyLED = OFF;                        // LED OFF
        wait(Dig);                          // Delay
    }
}
```

FIG. 8.88   Program listing.

8.26 PROJECT 22—SOUND LEVEL METER

8.26.1 Description

This is a sound level project. A small electret microphone-based preamplifier module is connected to analog input PA_0 of the Nucleo-F411RE development board. The project measures the peak-to-peak voltage output of the amplifier and displays this voltage on the PC screen in millivolts. This voltage is proportional to the ambient sound level (in dB) and should be calibrated for accuracy using a professional sound level meter.

8.26.2 Aim

The aim of this project is to show how a sound level meter can be designed.

8.26.3 Block Diagram

The block diagram of the project is shown in Fig. 8.89. In this project, the HiLetgo audio module (www.iletgo.com) is used. This module has an onboard electret microphone and a MAX4466-type preamplifier with adjustable gain.

8.26.4 Circuit Diagram

Fig. 8.90 shows the circuit diagram of the project. The HiLetgo audio module has three pins: OUT, GND, and VCC. The VCC and GND are connected to the +3.3 V and GND of the development board. The module can operate from a supply voltage +2.4 to +5 V. The OUT pin is connected to analog input PA_0. The output of the audio module is at the DC voltage VCC/2 when there is no sound input. A small potentiometer is provided at the bottom of the module so that the gain of the preamplifier can be varied from 25 to 125. You may have to adjust this potentiometer for better response.

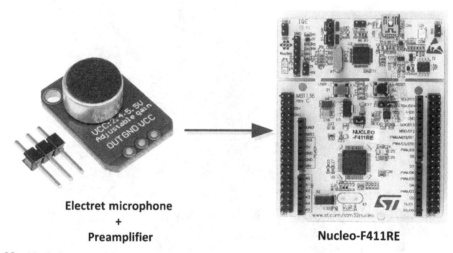

Electret microphone
+
Preamplifier

Nucleo-F411RE

FIG. 8.89   Block diagram of the project.

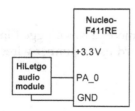

FIG. 8.90   Circuit diagram of the project.

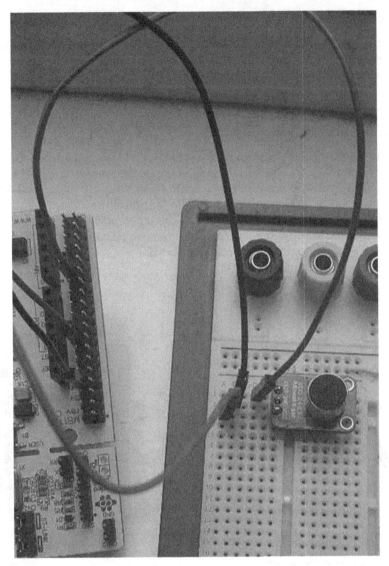

FIG. 8.91   The project constructed on a breadboard.

8.26.5 The Construction

As shown in Fig. 8.91, the audio module was plugged-in on a breadboard and connections were made to the development board using jumper wires.

8.26.6 The PDL

The PDL of the program is shown in Fig. 8.92.

BEGIN
>   Configure PA_0 as analog input
>   Configure Timer
>   **DO WHILE** Timeout < 50ms
>>   Read maximum signal value
>>   Read minimum signal value
>   **ENDDO**
>   Peak to peak signal = maximum value – minimum value
>   Display peak signal
>   Wait 1 second

END

FIG. 8.92   PDL of the program.

8.26.7 Program Listing

The program listing (program:**SoundLevel**) is shown in Fig. 8.93. In this program, the Timer is used to create a 50 ms (20 Hz) window and the maximum and minimum values of the audible sound signal are read inside this window. **tim.reset()** resets Timer to 0, **tim.start()** starts the Timer, and tim.stop() stops the timer. The Timer reading is read in milliseconds using the **tim.read_ms()** statement. The maximum and minimum values of the input signal are stored in floating point variables **SMax** and **SMin**, respectively, in millivolts. Floating point variable **Peak** stores the peak-to-peak value of the sound signal. This value is displayed at line 4, column 0 of the PC screen. The above process is repeated after 1s delay. The display shows the sound level reading in millivolts. This reading should be calibrated and converted into Decibels using a professional sound level meter. Fig. 8.94 shows an output from the program.

8.26.8 Suggestions for Additional Work

Modify the project by connecting eight LEDs to the development board. Then, modify the program in Fig. 8.93 so that the number of LEDs Turning ON and OFF depend on the sound level of the measured audio signal (just like the disco lights).

8.27 USING THE DIGITAL-TO-ANALOG CONVERTER

Digital-to-analog converter (DAC) modules are important parts of microcontrollers as they enable analog signals to be output from the microcontroller. DACs receive digital signals and convert them into analog form. For example, we can generate and send out various waveforms through the DAC module, or we can filter an analog signal and send out the filtered waveform through the DAC module.

```
/**********************************************************************
                        SOUND LEVEL METER
                        =================

This is a sound level meter project. An electret microphone with an
amplifier is connected to analog input PA_0 of the develomet board.
The peak-to-peak output voltage of the amplifie is measured and is
displayed on the PC screen in millivolts. This reading is proportional
to the ambient sound level and needs to be calibrated in dB.

Author: Dogan Ibrahim
Date  : August 2018
File  : SoundLevel
**********************************************************************/
#include "mbed.h"

Serial MyPC(USBTX, USBRX);
AnalogIn Sound(PA_0);
Timer tim;

// Clear the screen
//
void clrscr()
{
    char clrscr[] = {0x1B, '[', '2' , 'J',0};
    MyPC.printf(clrscr);
}


//
// Goto specified line and column
//
void gotoscr(int line, int column)
{
    char scr[] = {0x1B, '[', 0x00, ';' ,0x00, 'H', 0};
    scr[2] = line;
    scr[4] = column;
    MyPC.printf(scr);
}


int main()
{
    double mV;
    float SMax = 0,Peak,SMin = 4096;

//
// Read the peak-topeak output voltage of the audio amplifier,
// then display the voltage on the PC screen
//
    while(1)                                              // Do forever
```

FIG. 8.93 Program listing.

(Continued)

```
    {
        tim.reset();                              // Reset Timer
        tim.start();                              // Start Timer
        SMax = 0.0;
        SMin = 3300.0;

        while(tim.read_ms() < 50)                 // Do for 50ms
        {
            mV = 3300.0f * Sound.read();          // In mV
            if(mV > SMax)                         // Find Max
                SMax = mV;
            else if(mV < SMin)                    // Find Min
                SMin = mV;
        }

        Peak = SMax - SMin;                       // Peak-to-peak
        tim.stop();                               // Stop Timer
        clrscr();                                 // Clear screen
        gotoscr('4', '0');                        // Line 4, col 0
        MyPC.printf("Sound Level = %5.2f", Peak); // Display
        wait(1.0);                                 // Wait 1 second
    }
}
```

FIG. 8.93, CONT'D

Just like the ADCs, the resolution of a DAC depends on the number of bits used in the conversion process. Also, as with the ADCs, DACs also have reference voltages and the output analog voltage depends on the value of this reference voltage. For example, with a 12-bit (4096 steps) DAC and with a reference voltage of +3.3 V, each DAC step corresponds to $3300/4096 = 0.805$ mV. Thus, for example, the 12-bit digital value of "1011 0000 1111" (i.e., decimal 2831) corresponds to $2831 \times 0.805 = 2.278$ V.

The STM32F411RET6 processor on the Nucleo-F411RE development board has no built-in DAC modules. Most other Nucleo boards, however, have one or more DAC modules. For example, the Nucleo-L476RG development board has two built-in DAC modules. Because the DAC is an important part of a microcontroller, we shall see in this section how to use a DAC module by programming the Nucleo-L476RG board using Mbed.

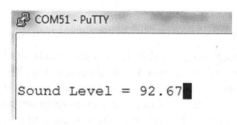

FIG. 8.94 An output from the program.

There are two 12-bit DACs on the Nucleo-L476RG development board, available at pins PA_4 and PA_5. A DAC port is configured using the statement AnalogOut. The following functions are available with the **AnalogOut**:

write: This function sends out analog data to the specified analog pin. Valid values are 0.0–1.0 where 0.0 corresponds to 0 V and 1.0 corresponds to +3.3 V.

write_u16: This function sends out analog data to the specified analog pin. Valid values are 0–65,535, where 0 corresponds to 0 V and 65,535 corresponds to +3.3 V.

read: This function returns the analog voltage sent to the specified analog pin. Returned values are 0.0–1.0.

As an example, analog output of 0.5 corresponds to $0.5 \times 3.3\,V = 1.65\,V$.

8.28 PROJECT 23—GENERATING FIXED VOLTAGE

8.28.1 Description

Perhaps the simplest application of a DAC is to generate fixed voltages. In this project, a function is created to generate fixed voltages. The main program calls this function to generate 0, 1, 2, and 3 V waveforms with 100 ms delay between each output. This process is repeated forever until stopped by the user.

8.28.2 Aim

The aim of this project is to show how the DAC module can be programmed on a Nucleo-L476RG development board.

8.28.3 Circuit Diagram

In this project, pin PA_4 (pin 32 on connector CN7) of the Nucleo-L476RG is connected to a digital oscilloscope so that the waveform can be displayed and recorded. In this project, the PCSGU250 Velleman PC-based oscilloscope is used.

8.28.4 The PDL

Fig. 8.95 shows the program PDL.

8.28.5 Program Listing

Before compiling the program, you should add the Nucleo-L476RG board to your compiler. The program listing (program: **FixedV**) is shown in Fig. 8.96. At the beginning of the program, statement **AnalogOut** is used to assign variable **aout** to DAC port PA_4. Function **GenerateV** receives the voltage to be generated as its argument and generates the required voltage. Inside the main program, the wile statement is used to create an endless loop. Inside this loop, 0, 1, 2, and 3 V are generated with 100 ms delay between each output.

BEGIN/MAIN
    Configure PortA_4 as DAC
    CALL GenerateV with 0V
    Wait 100ms
    CALL GenerateV with 1V
    Wait 100ms
    CALL GenerateV with 2V
    Wait 100ms
    CALL GenerateV with 3V
    Wait 100ms
END/MAIN

BEGIN/GenerateV (V)
    Generate output voltage with amplitude V
END/GenerateV

FIG. 8.95   Program PDL.

```
/****************************************************************************

                        GENERATE FIXED VOLTAGE
                        ======================

This is an example DAC program. In thsi program fixed voltages of 0V, 1V,
2V, and 3V are generated with 100ms delay between each output

Author: Dogan Ibrahim
Date  : August 2018
File  : FixedV
*****************************************************************************/
#include "mbed.h"

Serial MyPC(USBTX, USBRX);
AnalogOut aout(PA_4);

//
// This function generates V voltage at the DAC port
//
void GenerateV(float V)
{
    aout = V / 3.3f;
}


int main()
{
    while(1)                                // Do forever
    {
        GenerateV(0.0);                     // Generate 0V
        wait(0.1);                          // Wait 100ms
        GenerateV(1.0);                     // Generate 1V
        wait(0.1);                          // Wait 1000ms
        GenerateV(2.0);                     // Genearte 2V
        wait(0.1);                          // Wait 100ms
        GenerateV(3.0);                     // Generate 3V
        wait(0.1);                          // Wait 100ms
    }
}
```

FIG. 8.96   Program listing.

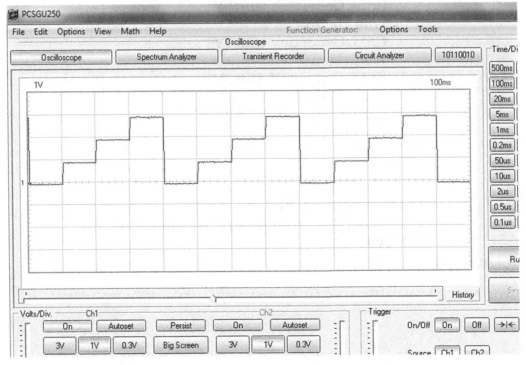

FIG. 8.97   Generated waveform.

Fig. 8.97 shows the generated staircase waveform. In this figure, the vertical voltage axis is 1 V/division, and the horizontal time axis is 100 ms/division.

8.29 PROJECT 24—SAWTOOTH WAVEFORM

8.29.1 Description

In this project, we generate a sawtooth waveform with 10 steps with a period of 20 ms, that is, the time delay between each step is 2 ms.

8.29.2 Aim

The aim of this project is to show how a sawtooth waveform can be generated.

8.29.3 Circuit Diagram

In this project, analog input PA_4 is used as in the previous project.

BEGIN
      Configure DAC port PA_4 as analog output
      DO FOREVER
            Send 10 steps to DAC port with 2ms each
      ENDDO
   END

FIG. 8.98   Program PDL.

8.29.4 The PDL

Fig. 8.98 shows the PDL of the program.

8.29.5 Program Listing

The program listing (program: **Sawtooth**) is shown in Fig. 8.99. At the beginning of the program, variable **aout** is configured as analog output and is assigned to DAC port PA_4.

```
/**********************************************************************

                    GENERATE SAWTOOTH WAVEFORM
                    ==========================

This program generates a sawtooth waveform with 10 rising steps where
each step width is 2ms. i.e. the period of the waveform is 20ms.

Author: Dogan Ibrahim
Date  : August 2018
File  : Sawtooth
***********************************************************************/
#include "mbed.h"

Serial MyPC(USBTX, USBRX);
AnalogOut aout(PA_4);

int main()
{
   float k;

   while(1)                                   // Do forever
   {
       for(k = 0.0f; k < 1.0f; k = k + 0.1f)   // Do 10 times
       {
           aout = k;                           // Analog out
           wait(0.002);                        // Wait 2ms
       }
   }
}
```

FIG. 8.99   Program listing.

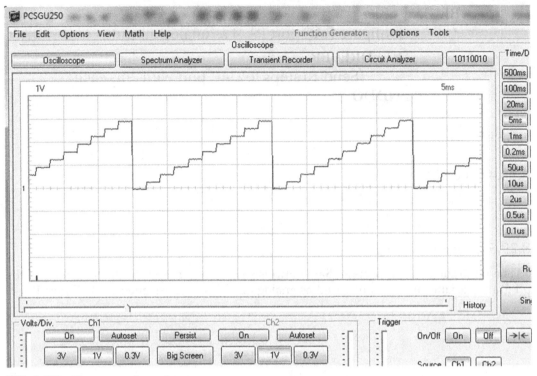

FIG. 8.100   The generated waveform.

The program consists of an endless loop. Inside this loop, a **for** loop is used to iterate 10 times. Inside this **for** loop, 10 steps are generated where the width of each step is 2 ms. Therefore, as shown in Fig. 8.100, the period of the generated sawtooth waveform is 20 ms. In this figure, the vertical axis is 1 V/division and the horizontal axis is 5 ms/division.

8.30 PROJECT 25—TRIANGULAR WAVEFORM

8.30.1 Description

In this project, we generate a triangular waveform with 100 steps going up and 100 steps coming down. The step width is set to 100 μs, that is, the period of the generated triangular waveform is 20 ms.

8.30.2 Aim

The aim of this project is to show how a triangular waveform can be generated.

BEGIN
      Configure DAC port PA_4 as analog output
      DO FOREVER
            Send 100 rising steps to DAC port with 100 microseconds each
            Send 100 falling steps to DAC port with 100 microseconds each
      ENDDO
END

FIG. 8.101   Program PDL.

8.30.3 Circuit Diagram

In this project, analog input PA_4 is used as in the previous project.

8.30.4 The PDL

Fig. 8.101 shows the PDL of the program.

8.30.5 Program Listing

The program listing (program: **Triangular**) is shown in Fig. 8.102. At the beginning of the program, variable **aout** is configured as analog output and is assigned to DAC port PA_4. The program consists of an endless loop. Inside this loop, two **for** loops are used where each one iterates 100 times with step size of 100 μs. One **for** loop is for the rising edge of the signal and the other **for** loop is for the falling edge of the signal. Therefore, as shown in Fig. 8.103, the period of the generated triangular waveform is 20 ms. This figure the vertical axis is 1 V/division, and the horizontal axis is 5 ms/division. Notice in this diagram that the period is not exactly 20 ms. This is because the wait statements are not very accurate. Also, the **for** statements add additional time to the loop and as a result the period of the waveform is slightly over 20 ms. We can reduce the delay by trial and error to make the period exactly 20 ms. This is shown in Fig. 8.104 where the delay inside the two for loops were 92 μs, that is, **wait_us(92.0)** instead of **wait_us(100.0).**

8.31 PROJECT 26—SINE WAVEFORM

8.31.1 Description

In this project, we generate a sine waveform with 100 steps, each step 100 μs, thus having a period of 10 ms (frequency of 100 kHz).

```
/*************************************************************************
                      GENERATE TRIANGULAR WAVEFORM
                      ============================

This program generates a triangular waveform with 100 steps rising and 100
steps falling, where each step width is 100 microseconds. i.e. the period
of the waveform is 20ms

Author: Dogan Ibrahim
Date  : August 2018
File  : Triangular
*************************************************************************/
#include "mbed.h"

Serial MyPC(USBTX, USBRX);
AnalogOut aout(PA_4);

int main()
{
    float k;

    while(1)                                    // Do forever
    {
        for(k = 0.0f; k < 1.0f; k = k + 0.01f)  // Do 100 times
        {
            aout = k;                           // Analog out
            wait_us(100.0);                     // Wait 100us
        }

        for(k = 1.0f; k > 0; k = k - 0.01f)     // Do 100 times
        {
            aout = k;                           // Analog out
            wait_us(100.0);                     // Wait 100us
        }
    }
}
```

FIG. 8.102   Program listing.

8.31.2 Aim

The aim of this project is to show how a sine waveform can be generated.

8.31.3 Circuit Diagram

In this project, analog input PA_4 is used as in the previous project.

8.31.4 The PDL

Fig. 8.105 shows the PDL of the program.

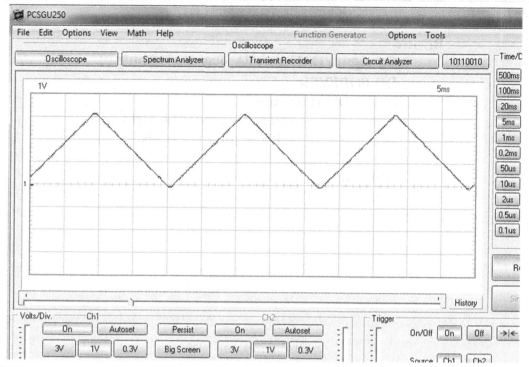

FIG. 8.103 The generated waveform.

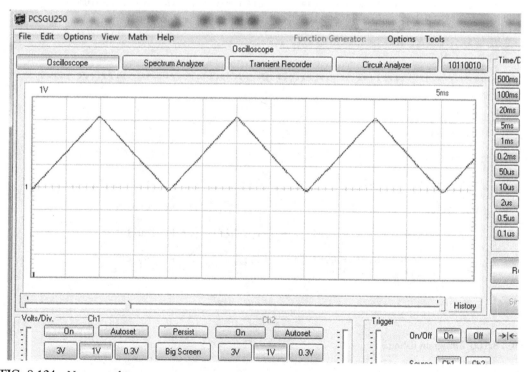

FIG. 8.104 New waveform.

BEGIN
 Configure DAC port PA_4 as analog output
 DO FOREVER
  DO 100 times
   Generate a sine waveform step
   Wait 100 microseconds
  ENDDO
 ENDDO
END

FIG. 8.105   Program PDL.

```
/*************************************************************************

                        GENERATE SINE WAVEFORM
                        ======================

This program generates a sine waveform with 100 steps, each step 100
microsseconds. The period of the generated waveform is therefore 10ms.
The waveform has the peak amplitude of +3.3V.

Author: Dogan Ibrahim
Date  : August 2018
File  : SineWave
*************************************************************************/
#include "mbed.h"

Serial MyPC(USBTX, USBRX);
AnalogOut aout(PA_4);

int main()
{
   float k, Pi = 3.14159;

   while(1)                                   // Do forever
   {
       for(k = 0.0f; k < 2.0f; k = k + 0.02f)  // Do 100 times
       {
           aout = 0.5f + 0.5f*sin(k*Pi);       // Analog out
           wait_us(100.0);                     // Wait 100us
       }

   }
}
```

FIG. 8.106   Program listing.

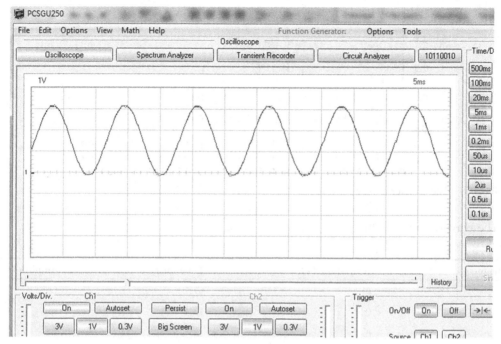

FIG. 8.107 The generated waveform.

8.31.5 Program Listing

The program listing (program: **SineWave**) is shown in Fig. 8.106. At the beginning of the program, variable **aout** is configured as analog output and is assigned to DAC port PA_4. The program consists of an endless loop. Inside this loop, a **for** loop is used which iterates 100 times with step size of 100 μs. Therefore, the period of the generated waveform is $100 \times 100\,\mu s = 10\,ms$. The amplitude of the waveform is shifted up so that it is always positive. The maximum and minimum values of the **sin** function are ± 1. Therefore, 0.5 ± 1 takes values between 0 and 1, that is, the amplitude of the generated waveform is between 0 and +3.3 V. Fig. 8.107 shows the generated sine waveform. In this figure, the vertical axis is 1 V/division and the horizontal axis is 5 ms/division.

8.32 PROJECT 27—ARBITRARY PERIODIC WAVEFORM

8.32.1 Description

In this project, we generate an arbitrary periodic waveform with a period of 20 ms where the details of the waveform are as shown in Fig. 8.108.

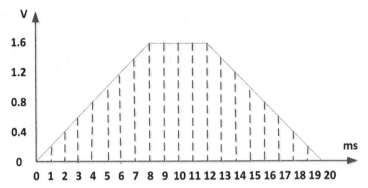

FIG. 8.108   The waveform to be generated.

The waveform takes the following values:

Time (ms)	Amplitude (V)	Time (ms)	Amplitude (V)
0	0	11	1.6
1	0.2	12	1.6
2	0.4	13	1.4
3	0.6	14	1.2
4	0.8	15	1.0
5	1.0	16	0.8
6	1.2	17	0.6
7	1.4	18	0.4
8	1.6	19	0.2
9	1.6	20	0.0
10	1.6		

8.32.2 Aim

The aim of this project is to show how an arbitrary waveform can be generated.

8.32.3 Circuit Diagram

In this project, analog input PA_4 is used as in the previous project.

8.32.4 The PDL

Fig. 8.109 shows the PDL of the program.

BEGIN/MAIN
  Configure DAC port PA_4 as analog output
  Store waveform voltage samples in array Waveform
  DO FOREVER
    DO for all sample points
      CALL GenerateV(sample)
      Wait 1 millisecond
    ENDDO
  ENDDO
END/MAIN

BEGIN/GenerateV(sample)
  Generate voltage at the DAC equivalent to sample
END/GenerateV

FIG. 8.109 Program PDL.

8.32.5 Program Listing

Fig. 8.110 shows the program listing (program: **Arbitrary**). At the beginning of the program, the waveform voltage sample points are stored in a floating point array called **Waveform**. Function **GenerateV** generates the given voltage at the DAC port. Inside the main program, the voltage sample points are sent to function **GenerateV** to generate the required waveform.

Fig. 8.111 shows the generated waveform. In this figure, the vertical axis is 0.3 V/division and the horizontal axis is 2 ms/division.

8.33 PROJECT 28—WAVEFORM GENERATOR

8.33.1 Description

This is a waveform generator program which can generate the following waveforms: fixed voltage, sawtooth waveform, triangular waveform, sine waveform, and arbitrary waveform. The parameters of each type of waveform are received from the keyboard.

8.33.2 Aim

The aim of this project is to show how a waveform generator program can be developed.

```
/*****************************************************************

                    ARBITRARY PERIODIC WAVEFORM
                    ===============================

In this program an arbitrary periodic waveform is generated with a period
of 20ms. The details of teh waveform are given in the text.

Author: Dogan Ibrahim
Date  : August 2018
File  : Arbitrary
*****************************************************************/
#include "mbed.h"

Serial MyPC(USBTX, USBRX);
AnalogOut aout(PA_4);

float Waveform[] = {0.0,0.2,0.4,0.6,0.8,1.0,1.2,1.4,1.6,1.6,1.6,1.6,1.6,
                    1.4,1.2,1.0,0.8,0.6,0.4,0.2,0.0};
//
// This function generates V voltage at the DAC port
//
void GenerateV(float V)
{
    aout = V / 3.3f;
}


int main()
{
    int k;

    while(1)                                // Do forever
    {
        for(k = 0; k <= 20; k++)            // Get waveform samples
        {
            GenerateV(Waveform[k]);         // Generate waveform
            wait_ms(1.0);                   // Wait 1ms
        }
    }
}
```

FIG. 8.110   Program listing.

8.33.3 Circuit Diagram

In this project, analog input PA_4 is used as in the previous project.

8.33.4 The PDL

Fig. 8.112 shows the PDL of the program.

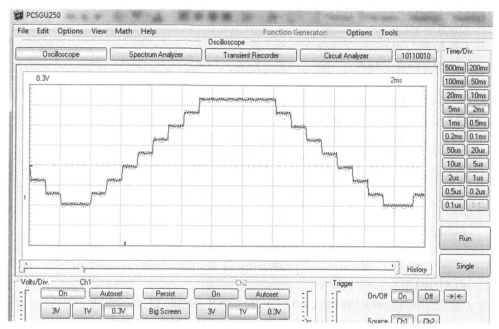

FIG. 8.111 Generated waveform.

8.33.5 Program Listing

Fig. 8.113 shows the program listing (program: **Waveform**). The program is MENU based and consists of a number of functions. The MENU has five options and enables users to choose the type of waveform required. After the MENU choice, a switch statement is used to read the required parameters for each waveform. Option 1 generates a fixed voltage at the DAC port. Calling function **GenerateFixed** generates the required fixed voltage at the DAC port. Option 2 generates a sawtooth waveform. The number of steps required in the waveform, the period of the waveform, and its amplitude are all read from the keyboard. Calling function **GenerateSawtooth** generates the required sawtooth waveform. Option 3 generates a triangular waveform with same parameters as the sawtooth waveform. Calling function **GenerateTriangular** generates the required triangular waveform. Option 4 generates a sine wave with the required peak-to-peak amplitude, number of points, and the period. Calling function **GenerateSine** generates the required sine waveform. Finally, option 5 generates any kind of waveforms where the sample time (in ms) and the sample points (in volts) are entered by the user. Calling function **GenerateArbitrary** generates the required waveform. After a waveform is generated, the program remains at the requested function. Pressing the Reset button on the development board will restart the program so that a new waveform can be generated if desired.

```
BEGIN/MAIN
        Configure PA_4 as analog output
        Clear screen and home cursor
        Display MENU
        Get MENU choice
        IF choice = 1 THEN
                Read required voltage
                CALL GenerateFixed (V)
        ELSE IF Choice = 2 THEN
                Read required amplitude, steps, and period
                CALL GenerateSawtooth (amplitude, steps, period)
        ELSE IF Choice = 3 THEN
                Read required amplitude, steps, and period
                CALL GenerateTriangular (amplitude, steps, period)
        ELSE IF Choice = 4 THEN
                Read required peak-to-peak amplitude, steps, and period
                CALL GenerateSine (peak-to-peak amplitude, steps, period)
        ELSE IF Choice = 5 THEN
                Read time interval, no of points, and sample point values
                CALL GenerateArbitrary (time interval, points, sample point values)
        ENDIF
END/MAIN


BEGIN/GenerateFixed (V)
        Generate fixed voltage V
END/GenerateFixed

BEGIN/GenerateSawtooth(amplitude, steps, period)
        Genearte sawtooth waveform with required parameters
END/GenerateSawtooth

BEGIN/GenerateTriangular(amplitude, steps, period)
        Genearte triangular waveform with required parameters
END/GenerateTriangular

BEGIN/GenerateSine(peak-to-peak amplitude, steps, period)
        Genearte sine waveform with required parameters
END/GenerateSine

BEGIN/GenerateArbitray(time interval, no of points, sample point values)
        Genearte waveform with required time interval,no of points, and sample point values
END/GenerateArbitrary
```

FIG. 8.112   Program PDL.

```
/***********************************************************************

                    WAVEFORM GENERATOR
                    ==================

This program generates fixed voltage, sawtooth waveform, triangular
waveform, sine waveform, and arbitrary waveform. The parameters of
each type of waveform are entered from the keyboard

Author: Dogan Ibrahim
Date  : August 2018
File  : Waveform
***********************************************************************/
#include "mbed.h"

Serial MyPC(USBTX, USBRX);
AnalogOut aout(PA_4);

//
// Clear the screen
//
void clrscr()
{
    char clrscr[] = {0x1B, '[', '2' , 'J',0};
    MyPC.printf(clrscr);
}

//
// Home the cursor
//
void homescr()
{
    char homescr[] = {0x1B, '[' , 'H' , 0};
    MyPC.printf(homescr);
}


//
// This function generates fixed voltage at the DAC port
//
void GenerateFixed(float V)
{
    aout = V / 3.3f;
}

//
// This function generates a Sawtooth waveform. The required number
// of staps and the period are arguments of the function
//
void GenerateSawtooth(float Amp, int Steps, float Period)
{
    float k, amplitude, inc, dly;

    amplitude = Amp / 3.3f;                            // Required amplitude
```

FIG. 8.113 Program listing.

(Continued)

```
    inc = amplitude / Steps;
    dly = Period / Steps;
    dly = dly / 1000.0f;                     // in seconds

    while(1)                                 // Do forever
    {
        for(k = 0.0f; k <= amplitude; k = k + inc)
        {
            aout = k;                        // Analog out
            wait(dly);                       // Wait dly sec
        }
    }
}


//
// This function generates a Triangular waveform. The required number
// of staps and the period are arguments of the function
//
void GenerateTriangular(float Amp, int Steps, float Period)
{
    float k, amplitude, inc, dly;

    amplitude = Amp / 3.3f;                  // Required amplitude
    inc = amplitude / Steps;
    dly = Period / Steps / 2;
    dly = dly / 1000.0f;                     // in seconds

    while(1)                                 // Do forever
    {
        for(k = 0.0f; k <= amplitude; k = k + inc)  // Rising
        {
            aout = k;                        // Analog out
            wait(dly);                       // Wait dly sec
        }

        for(k = amplitude; k > 0; k = k - inc)      // Falling
        {
            aout = k;                        // Analog out
            wait(dly);                       // Wait dly sec
        }
    }

}


//
// This function generates a Sine waveform. The required number
// of staps and the period are arguments of the function
//
void GenerateSine(float Amp, int Steps, float Period)
{
    float k, amplitude, inc, dly;
    float Pi = 3.14159;
```

FIG. 8.113, CONT'D

(Continued)

```
    amplitude = Amp / 3.3f / 2;                      // Required amplitude
    inc = 2.0 / Steps;
    dly = Period / Steps;
    dly = dly / 1000.0f;                             // in seconds

    while(1)                                         // Do forever
    {
        for(k = 0.0f; k < 2.0f; k = k + inc)
        {
            aout = amplitude + amplitude*sin(k*Pi); // Analog out
            wait(dly);                               // Wait dly sec
        }

    }
}


//
// This function generates a arbitray waveform. The time interval,
// number of points and the waveform array are the arguments
//
void GenerateArbitrary(float T, int P, float W[])
{
    float Tim = T / 1000.0f;                         // In seconds

    while(1)                                         // Do forever
    {
        for(int k = 0; k <= P; k++)                  // Get waveform samples
        {
            GenerateFixed(W[k]);                     // Generate waveform
            wait(Tim);                               // Wait in secs
        }
    }
}


//
// Start of MAIN program. The main program displays a MENU for the user
// to choose what type of waveform is required. Then the parameters of
// teh chosen waveform are read from the keyboard
//
int main()
{
    int choice, RequiredSteps, Points;
    float RequiredVoltage, Amplitude, Period, TimeInterval;
    float Waveform[50];
//
// Display MENU of options
//
    clrscr();
    homescr();
    MyPC.printf("\n\rWAVEFORM GENERATOR");
    MyPC.printf("\n\r==================");
    MyPC.printf("\n\r");
    MyPC.printf("\n\r1. Generate Fixed Voltage");
```

FIG. 8.113,   CONT'D

(Continued)

```
    MyPC.printf("\n\r2. Generate Sawtooth Waveform");
    MyPC.printf("\n\r3. Generate Triangular Waveform");
    MyPC.printf("\n\r4. Generate Sine Waveform");
    MyPC.printf("\n\r5. Generate Arbitrary Waveform");
    MyPC.printf("\n\r");
    MyPC.printf("Choice: ");
    MyPC.scanf("%d", &choice);
    MyPC.printf("\n\r");

//
// Action depending on the MENU choice. Case 1: Fixed voltage, Case:2
// Sawtooth waveform, Case 3: Triangular waveform, 4: Sine waveform,
// Case 5: Arbitrary waveform
//
    switch (choice)
    {
        case 1:
                MyPC.printf("\n\r\n\rRequired Fixed Voltage (1 - 3.3V): ");
                MyPC.scanf("%f", &RequiredVoltage);
                MyPC.printf("\n\rGenerating %f Volt", RequiredVoltage);
                GenerateFixed(RequiredVoltage);
                break;
        case 2:
                MyPC.printf("\n\r\n\rEnter No of Steps: ");
                MyPC.scanf("%d", &RequiredSteps);
                MyPC.printf("\n\rEnter Period (ms): ");
                MyPC.scanf("%f", &Period);
                MyPC.printf("\n\rEnter Amplitude (0 - 3.3V): ");
                MyPC.scanf("%f", &Amplitude);
                MyPC.printf("\n\rGenerating Sawtooth Waveform");
                GenerateSawtooth(Amplitude, RequiredSteps, Period);
                break;
        case 3:
                MyPC.printf("\n\rEnter No of Steps: ");
                MyPC.scanf("%d", &RequiredSteps);
                MyPC.printf("\n\rEnter Period (ms): ");
                MyPC.scanf("%f", &Period);
                MyPC.printf("\n\rEnter Amplitude (0 - 3.3V): ");
                MyPC.scanf("%f", &Amplitude);
                MyPC.printf("\n\rGenerating Triangular Waveform");
                GenerateTriangular(Amplitude, RequiredSteps, Period);
                break;
        case 4:
                MyPC.printf("\n\rEnter P-P Amplitude (1 - 3.3V): ");
                MyPC.scanf("%f", &Amplitude);
                MyPC.printf("\n\rEnter No of Steps: ");
                MyPC.scanf("%d", &RequiredSteps);
                MyPC.printf("\n\rEnter Period (ms): ");
                MyPC.scanf("%f", &Period);
                MyPC.printf("\n\rGenerating Sine Waveform");
                GenerateSine(Amplitude, RequiredSteps, Period);
                break;
```

FIG. 8.113,   CONT'D

(Continued)

```
case 5:
        MyPC.printf("\n\rEnter Time Interval (ms): ");
        MyPC.scanf("%f", &TimeInterval);
        MyPC.printf("\n\rEnter Amplitude pairs");
        MyPC.printf("\n\rHow Many Point Are There?: ");
        MyPC.scanf("%d", &Points);
        MyPC.printf("\n\r\n\rEnter Amplitude Values for Every ms: ");
        for(int k = 0; k < Points; k++)
        {
            MyPC.printf("\n\rAmplitude Value for %d ms: ", k);
            MyPC.scanf("%f", &Waveform[k]);
        }
        MyPC.printf("\n\rGenerating Arbitrary Waveform");
        GenerateArbitrary(TimeInterval, Points, Waveform);
        break;
    }
}
```

FIG. 8.113,   CONT'D

Figs. 8.114 and 8.115 show some runs of the program. In the first figure, a triangular waveform is generated. The second figure generates an arbitrary waveform with the given specifications.

It should be noted that the period of the generated waveforms become less accurate as the frequency increases. This is because of the overheads in the program such as the delays caused by the **for** loops and inaccuracies of the **wait** statements. More accurate results could be obtained if the waveforms are generated inside timer ISRs.

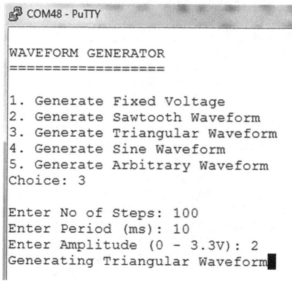

FIG. 8.114   Generating a triangular waveform.

8.33.6 Suggestions for Additional Work

Modify the program given in Fig. 8.99 by moving the code inside a timer ISR.

8.34 USING THE PWM

In many microcontroller-based applications, we may want to drive loads which may require large currents or voltages. Some examples are large motors, actuators, heaters, etc. The output current and voltage capabilities of microcontrollers are very limited where the maximum output current capacity is around 20 mA and the maximum output voltage is +3.3 or +5 V depending on the type of microcontroller used. PWM is used to drive loads that require large currents or voltages.

Basically, PWM is a positive going square waveform as shown in Fig. 8.116. The waveform has two parts: On time where the pulse if present, and the OFF time where there is no pulse. By varying the ratio of the ON time to OFF time, we can effectively vary the average DC voltage applied to the load. The PWM waveform usually drives the load through a switching power transistor or a transducer.

```
WAVEFORM GENERATOR
===================

1. Generate Fixed Voltage
2. Generate Sawtooth Waveform
3. Generate Triangular Waveform
4. Generate Sine Waveform
5. Generate Arbitrary Waveform
Choice: 5

Enter Time Interval (ms): 1
Enter Amplitude pairs
How Many Point Are There?: 10

Enter Amplitude Values for Every ms:
Amplitude Value for 0 ms: 0.1
Amplitude Value for 1 ms: 0.2
Amplitude Value for 2 ms: 0.3
Amplitude Value for 3 ms: 0.4
Amplitude Value for 4 ms: 0.5
Amplitude Value for 5 ms: 0.4
Amplitude Value for 6 ms: 0.3
Amplitude Value for 7 ms: 0.2
Amplitude Value for 8 ms: 0.1
Amplitude Value for 9 ms: 0
Generating Arbitrary Waveform
```

FIG. 8.115   Generating an arbitrary waveform.

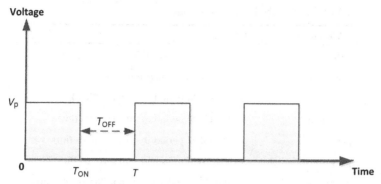

FIG. 8.116   PWM waveform.

The period T of the PWM waveform is the sum of its ON and OFF times, that is, $T = T_{ON} + T_{OFF}$. The duty cycle (D) of the waveform is the ratio of the ON time to its period, that is,

$$D = T_{ON}/T$$

The duty cycle is usually expressed as a percentage and it can vary between 0% and 100%. About 0% corresponds to the case where there is no ON time and 100% corresponds to the case where there is no OFF time, that is,

$$D = (T_{ON}/T_{OFF}) \times 100\%$$

It can be shown that the average voltage supplied to a load is given by

$$\text{Average} = D \times V_p$$

where V_p is the peak value of the PWM waveform. Therefore, by varying the duty cycle between 0% and 100%, we can effectively vary the average voltage applied to the load. In the case of the Nucleo-F411RE development board, the peak output voltage is +3.3V and the above equation reduces to

$$\text{Average} = 3.3 \times D$$

At 100% duty cycle the average voltage is +3.3V, at 50% duty cycle the average voltage is 1.65V, and at 0% duty cycle the average voltage is 0V.

Mbed PWM Functions

Mbed provides several functions for generating PWM waveforms. These functions are shown in Table 8.4.

Nucleo-F411RE Development Board PWM Channels

The Nucleo-F411RE development board supports large number of PWM channels. A list of the available channels is given below. In the list below, the first number is the Timer number controlling the PWM, and the second number is the PWM channel number. It is important to realize that different PWM channels on the same timers have shared periods. Therefore, changing one period will change others:

TABLE 8.4   Mbed PWM Functions

PWM Function	Description
write	Set the output duty-cycle
read	Read the current output duty cycle
period	Set the PWM period in seconds (float)
period_ms	Set the PWM period in milliseconds (int)
period_us	Set the PWM period in microseconds (int)
pulsewidth	Set the PWM pulsewidth in seconds (float)
pulsewidth_ms	Set the PWM pulsewidth in milliseconds (int)
pulsewidth_us	Set the PWM pulsewidth in microseconds (int)

PWM	GPIO Pin
PWM1/1	PA_8
PWM1/1N	PB_13, PA_7
PWM1/2	PA_9
PWM1/2N	PB_0, PB_14
PWM1/3	PA_10
PWM1/3N	PB_15, PB_1
PWM1/4	PA_11
PWM2/1	PA_0, PA_15, PA_5
PWM2/2	PA_1, PB_3
PWM2/3	PB_10
PWM3/1	PB_4, PA_6, PC_6
PWM3/2	PB_5, PC_7
PWM3/3	PC_8
PWM3/4	PC_9
PWM4/1	PB_6
PWM4/2	PB_7
PWM4/3	PB_8
PWM4/4	PB_9

Some PWM-based projects are given in the next few sections to show how the PWM functions can be used in projects. Notice that all these projects use the Nucleo-F411RE development board.

8.35 PROJECT 29—MELODY MAKER

8.35.1 Description

This project shows how sound with different frequencies can be generated with PWM waveforms using a simple buzzer device. The project shows how the simple melody **Happy Birthday** can be played using a buzzer.

8.35.2 Aim

The aim of this project is to show how PWM can programmed on the Nucleo-F411RE development board to generate sound with different frequencies.

8.35.3 Block Diagram

Fig. 8.117 shows the block diagram of the project.

8.35.4 Circuit Diagram

A buzzer is a small piezoelectric device that gives sound output when excited. Normally, buzzers are excited using square wave signals, also called pulse width modulated (PWM) signals. The frequency of the signal determines the pitch of the generated sound, and duty cycle of the signal can be used to increase or decrease the volume. Most buzzers operate in the frequency range 2—4 kHz. There are two types of buzzers: active and passive. Active buzzers operate with logic signals and they give a constant frequency sound. Passive buzzers require an audible waveform to generate sound. In this project,

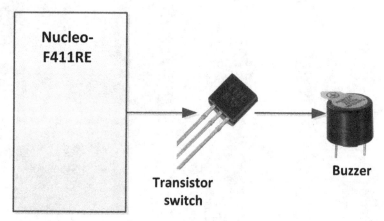

FIG. 8.117   Block diagram of the project.

a passive buzzer is used. The buzzer is connected to pin PWM port PA_8 (PWM1/1, pin 23 connector CN10) of the Nucleo-F411RE development board through a transistor switch as shown in Fig. 8.118.

8.35.5 The Construction

Fig. 8.119 shows the project constructed on a breadboard. Jumper wires are used to connect the buzzer, transistor, and the resistor to the development board.

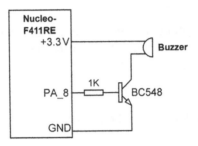

FIG. 8.118   Circuit diagram of the project.

FIG. 8.119   Project constructed on a breadboard.

8.35.6 The PDL

When playing a melody, each note is played for certain duration and with a certain frequency. In addition, a certain gap is necessary between two successive notes. The frequencies of the musical notes starting from middle C (i.e., C4) are given below. The harmonic of a note is obtained by doubling the frequency. For example, the frequency of C5 is $2 \times 262 = 524\,\text{Hz}$.

Notes	C4	C4#	D4	D4#	E4	F4	F4#	G4	G4#	A4	A4#	B4
Hz	261.63	277.18	293.66	311.13	329.63	349.23	370	392	415.3	440	466.16	493.88

In order to play the tune of a melody, we need to know its musical notes. Each note is played for certain duration and there is a certain time gap between two successive notes. The next thing we want is to know how to generate a sound with a required frequency and duration. In this project, we will be generating the classic **Happy Birthday** melody and thus we need to know the notes and their durations. These are given in the table below where the durations are in units of 400 ms (i.e., the values given in the table should be multiplied by 400 to give the actual durations in milliseconds).

Note	C4	C4	D4	C4	F4	E4	C4	C4	D4	C4	G4	F4	C4	C4	C5	A4	F4	E4	D4	A4#	A4#	A4	F4	G4	F4
Duration	1	1	2	2	2	3	1	1	2	2	2	3	1	1	2	2	2	2	2	1	1	2	2	2	4

The PDL of the project is shown in Fig. 8.120. Basically, two arrays are used to store the notes and their corresponding durations.

8.35.7 Program Listing

The program listing (program: **Melody**) is shown in Fig. 8.121. At the beginning of the program, PWM port PA_8 is assigned to variable **pwm** and the frequencies and durations of the melody are stored in two arrays called **Frequency** and **Duration,** respectively. Array **Period** stores the periods of the notes (the PWM function **pwm.period** requires the periods rather than the frequencies to be entered), array **Durations** stores the duration of each note in seconds (the PWM function **pwm.pulsewidth** requires the pulse width to be in seconds and not in milliseconds), and array **DutyCycle** stores the **Duty Cycle** of each PWM note as 50% (i.e., Period/2). Before the main program loop, the periods of the waveforms, duty cycles, and their durations are calculated so that the main program loop does not have to spend any time to do these calculations. Inside the program loop, the melody frequencies are generated with the required durations. Notice that the PWM waveform is stopped by setting its **pulsewidth** to 0. A small delay (100 ms) is introduced between each note. The melody is repeated after 3 s of delay.

BEGIN
      Configure PA_8 as PWM port
      Define melody frequencies
      Define melody durations
      DO for all Notes
            Calculate melody periods
            Calculate melody durations
            Set the Duty Cycles to 50%
      ENDDO
      DO FOREVER
            DO for all Notes
                  Set PWM period for a note
                  Set the pulse width for a note
                  Wait for the note duration
                  Stop the PWM
                  Wait 100ms note gap
                  Get next note
            ENDDO
            Wait 3 seconds
      ENDDO
END

FIG. 8.120   PDL of the project.

8.35.8 Suggestions for Additional Work

Modify the program given in Fig. 8.121 by changing the durations between the notes and see its effects. How can you make the melody run quicker?

8.36 PROJECT 30—ELECTRONIC ORGAN

8.36.1 Description

This is an electronic organ project. Keys **a–k** on the keyboard are used to generate the musical notes in the octave range **C5—C6**. Pressing a key generates the musical note for the duration of 200 ms. PWM waveforms with 50% duty cycles are generated for the notes. Serial interrupts are used in this program and all the notes are generated inside the serial ISR.

```
/***********************************************************************

                          MELODY MAKER
                          ============

In this project a buzzer is connected to PWM port PA_8. The program plays
the melody HAPPY BIRTHDAY. Array Frequency stores the frequencies of the
notes, array Period stores the periods of the notes, array Duration
stores the durations of each note in units of 400ms. Array Durations
stores the duration of each note in seconds, array DutyCycle stores the
Duty Cycle of each PWM note as 50% (i.e. Period / 2)

Author: Dogan Ibrahim
Date   : September 2018
File   : Melody
***********************************************************************/
#include "mbed.h"

#define MaxNotes 25
PwmOut pwm(PA_8);

int Frequency[MaxNotes] = {262,262,294,262,349,330,262,262,294,262,
                           392,349,262,262,524,440,349,330,294,466,
                           466,440,349,392,349};

int Duration[MaxNotes] = {1,1,2,2,2,3,1,1,2,2,2,3,1,1,2,2,2,2,
                          2,1,1,2,2,2,3};
float Period[MaxNotes];
float DutyCycle[MaxNotes];
float Durations[MaxNotes];

int main()
{

//
// Calculate the periods (in seconds), Duty Cycles (50%), and
// Durations (in seconds)before entering the loop
//
    for(int k = 0; k < MaxNotes; k++)
    {
        Period[k] = 1.0f / Frequency[k];
        DutyCycle[k] = Period[k] / 2.0f;
        Durations[k] = 400 * Duration[k] / 1000.0;
    }

//
```

FIG. 8.121 Program listing.

(Continued)

```
// Play the melody
//
    while(1)                                       // Do forever
    {
        for(int k = 0; k < MaxNotes; k++)          // Do for all notes
        {
            pwm.period(Period[k]);
            pwm.pulsewidth(DutyCycle[k]);          // 50% Duty Cycle
            wait(Durations[k]);                    // In seconds

            pwm.pulsewidth(0.0);                   // Stop PWM
            wait_ms(100);                          // 100ms Note gap
        }

        wait(3.0);                                 // Repeat after 3s
    }
}
```

FIG. 8.121, CONT'D

8.36.2 Aim

The aim of this project is to show how PWM can be programmed to create a simple electronic organ with one octave.

8.36.3 Block Diagram

The block diagram of the project is shown in Fig. 8.122.

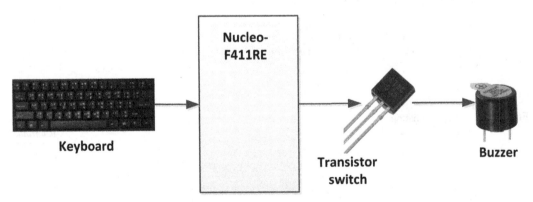

FIG. 8.122   Block diagram of the project.

8.36.4 Circuit Diagram

The circuit diagram of the project is same as in Fig. 8.118 where the buzzer is connected to PWM port PA_8 (PWM1/1, pin 23 connector CN10) of the Nucleo-F411RE development board through a transistor switch.

8.36.5 The PDL

The frequencies of the musical notes starting from C5 are given below (sharps and flats are not shown here). The harmonic of a note is obtained by doubling the frequency. For example, the frequency of C6 is $2 \times 523.25 = 1046.50\,Hz$.

Notes	C5	D5	E5	F5	G5	A5	B5	C6
Hz	523.25	587.33	659.25	698.46	783.99	880.00	987.77	1046.50

The PDL of the project is shown in Fig. 8.123.

BEGIN/MAIN
    Configure PA_8 as PWM port
    Define 8 musical note frequencies (C5 to C6)
    DO for all 8 Notes
        Calculate musical note periods and store in array Periods
        Set Duty Cycles to 50% in array DutyCycle
    ENDDO
    DO WHILE
        Wait here forever
    ENDDO
END/MAIN

BEGIN/ISR
    Read the key entered
    Generate a number 0 to 7 for keys a to k respectively
    Use this number to index array Periods
    Start PWM with selected period
    Set the Duty Cycle to 50%
    Wait 200ms
    Stop the PWM
END/ISR

FIG. 8.123   PDL of the project.

8.36.6 Program Listing

The program listing (program: **EOrgan**) is shown in Fig. 8.124. At the beginning of the program, PWM port PA_8 is assigned to variable **pwm** and the frequencies of the musical notes **C5–C6** are stored in array called **Notes**. Inside the main program, the serial interrupt is attached to function called **ISR** so that when a key is pressed on the keyboard the program jumps to this function. Also, the periods of the musical notes are calculated and stored in array called **Periods**. All the duty cycles are set to 50%. The main program then waits and does nothing else. Inside the ISR the characters typed on the keyboard are read and a **switch** statement is used to find out which character has been typed. Characters **a–k** are given numbers 0–7, respectively. These numbers are used to index array **Periods** to set the PWM period. A note is played for 200 ms and after this time the PWM is stopped, ready for the next key to be accepted.

8.36.7 Suggestions for Additional Work

Modify the program given in Fig. 8.124 by changing the durations a note is played.

Buzzers do not generate clear sounds. Modify the circuit diagram by using an audio amplifier (e.g., LM386) and a speaker (e.g., 8 ohm) to generate loud and clear musical tones.

8.37 PROJECT 31—VARYING THE LED BRIGHTNESS

8.37.1 Description

In this project, we vary the brightness of the User LED on the development board by sending a PWM waveform to the LED and then varying the duty cycle of this waveform so that the average voltage across the LED varies in relation to the duty cycle. The frequency of the PWM waveform is set to 50 Hz and its duty cycle is incremented in 10 steps every second, that is, the brightness is increased every 100 ms. When full brightness is reached, it is decreased again in 10 steps every second until the LED if turned OFF. This process is repeated continuously.

8.37.2 Aim

The aim of this project is to show how PWM can be programmed to vary the brightness of an LED.

8.37.3 Project PDL

If we turn the LED ON and OFF slowly, then we see a flashing LED. If on the other hand we shorten the ON and OFF times and turn ON for 50% and OFF for 50%, the LED will appear half as bright since half of the average voltage is applied to the LED. The important factor here is the duration. Therefore, if the ON and OFF times are very short, then the viewer will not notice the flashing effect and will see the LED dimmed. In general, if we

```
/*************************************************************************

                         ELECTRONIC ORGAN
                         ================

In this project a buzzer is connected to PWM port PA_8. The program is an
electronic organ that can play the notes for one octave. The following keys
are used on the keyboard for the notes:

a:C5, s:D5 d:E5 f:F5 g:G5 h:A5 j:B5 k:C6

Pressing a key plays the corresponding note for 200ms.

Author: Dogan Ibrahim
Date  : September 2018
File  : EOrgan
*************************************************************************/
#include "mbed.h"
Serial MyPC(USBTX, USBRX);

PwmOut pwm(PA_8);
#define MaxNotes 8

//
// Define the musical note (C5 - C6) frequencies
//
float Notes[MaxNotes] = {523.25,587.33,659.25,698.46,783.99,880.0,
                         987.77,1046.5};

float Periods[MaxNotes];
float DutyCycle[MaxNotes];

//
// Serial interrupt service routine. The program jumps here when
// a key is pressed on the keyboard
//
void ISR()
{
    char c;
    int k;

    c=MyPC.getc();                          // Get a key from keyboard

    switch (c)
    {
        case 'a':                           // Is it 'a' ?
            k = 0;
            break;
        case 's':                           // Is it 's' ?
            k = 1;
            break;
        case 'd':                           // Is it 'd' ?
            k = 2;
            break;
        case 'f':                           // Is it 'f' ?
```

FIG. 8.124   Program listing.

(Continued)

```
            k = 3;
            break;
        case 'g':                              // Is it 'g' ?
            k = 4;
            break;
        case 'h':                              // Is it 'h' ?
            k = 5;
            break;
        case 'j':                              // Is it 'j' ?
            k = 6;
            break;
        case 'k':                              // Is it 'k' ?
            k = 7;
            break;
    }

    pwm.period(Periods[k]);                    // Start the PWM
    pwm.pulsewidth(DutyCycle[k]);              // 50% Duty Cycle
    wait_ms(200);                              // Play for 200ms
    pwm.pulsewidth(0.0);                       // Stop PWM
}


int main()
{
    MyPC.attach(&ISR, Serial::RxIrq);          // Attach Serial Int

//
// Calculate the periods (in seconds), Duty Cycles (50%) before
// entering the loop
//
    for(int k = 0; k < MaxNotes; k++)
    {
        Periods[k] = 1.0f / Notes[k];
        DutyCycle[k] = Periods[k] / 2.0f;
    }

    while(1)                                    // Wait for ISR
    {
    }
}
```

FIG. 8.124, CONT'D

use a PWM waveform with the frequency set to about 50 Hz (period = 20 ms), then the viewer will not notice the flashing effect. As the duty cycle is varied, the brightness of the LED will change. For example, a higher duty cycle will result in brighter light since the average voltage is higher. Similarly, a lower duty cycle will result in dimmer light. The PDL of the project is shown in Fig. 8.125.

BEGIN
> Configure User LED as PWM port
> Set the PWM frequency to 50Hz
> Period = 1 / frequency
> Set variable inc to Period / 10
> **DO** 10 times
>> Increment the Duty Cycle by inc
>> Wait 1 second
> **ENDDO**
> **DO** 10 times
>> Decrement the Duty Cycle by inc
>> Wait 1 second
> **ENDDO**
END

FIG. 8.125   PDL of the project.

8.37.4 Program Listing

Fig. 8.126 shows the program listing (program: **Dimmer**). At the beginning of the program, User LED (LED1) is configured as a PWM port, the PWM frequency is set to 50 Hz and its duty cycle is set to 0 to start with. Variable **inc** is set to one-tenth of the period and the duty cycle is incremented by this amount inside a **for** loop which iterates 10 times. Therefore, the brightness of the LED is incremented at every 100 ms. After reaching the full brightness (100% duty cycle), the duty cycle is decremented in 10 steps until the LED is turned OFF. This process is repeated forever. As a result, the LED seems to be getting brighter, and then dimmer.

8.38 SUMMARY

In this chapter, we have learned about the following:

- Using 7-segment displays
- PC serial interface
- Timer interrupts
- Analog to digital converter
- Digital to analog converter
- Waveform generation
- Generating sound
- Pulse width modulation

```
/*************************************************************************

                          LED DIMMER
                          ==========

In this project the brightness of the User LED on the Nucleo F411RE
development board is varied. A PWM waveform with the 50Hz frequency
is sent to the LED. The Duty Cycle of this waveform is increased in 10
increments in a second. i.e. every 100ms so that the brightness of the
LED increases. When full brightness is reached, the Duty Cycle is
decremented in 10 steps in a second so that the LED becomes dimmer.

Author: Dogan Ibrahim
Date  : September 2018
File  : Dimmer
*************************************************************************/
#include "mbed.h"

Serial MyPC(USBTX, USBRX);

PwmOut pwm(LED1);

float frequency = 50.0;                    // Frequency = 50Hz
float period = 1.0f / frequency;           // Period
float DutyCycle,inc;

int main()
{
    pwm.period(period);                    // Start the PWM
    inc = period / 10.0f;                  // Increment
    DutyCycle = 0.0;                       // Starting Duty Cycle

    while(1)                               // Do Forever
    {
        for(int k = 0; k < 10; k++)        // Do 10 times
        {
            pwm.pulsewidth(DutyCycle);     // Set Duty Cycle
            DutyCycle = DutyCycle + inc;   // Increment Duty Cycle
            wait(1.0);                     // Wait 1 second
        }

        for(int k = 10; k > 0; k--)        // Do 10 times
        {
            pwm.pulsewidth(DutyCycle);     // Set Duty Cycle
            DutyCycle = DutyCycle - inc;   // Decrement Duty Cycle
            wait(1.0);                     // Wait 1 second
        }
    }
}
```

FIG. 8.126   Program listing.

8.39 EXERCISES

1. Explain how a multiplexed 7-segment display works.
2. Some alphabetic characters can be displayed using a 7-segment display. Show with examples which characters can be displayed.
3. A two-digit 7-segment display is connected to a Nucleo-F411RE development board. In addition, two external buttons called UP and DOWN are connected to the development board. Write a program that will count up when button UP is pressed, and it will count down when button DOWN is pressed.
4. An LM35-type analog temperature sensor is connected to one of the analog inputs of a Nucleo-F411RE development board. In addition, a two-digit 7-segment display is connected to the development board. Write a program to display the temperature every 10 s.
5. Write a program to display a sine wave with a frequency of 1.5 kHz and a peak-to-peak value of 1.2 V.

Motor Control Projects

9.1 OVERVIEW

In this chapter we shall be developing projects using electric motors with the Nucleo-F411RE development board. The projects will cover the following types of electric motors: brushed DC (BDC) motors, servo motors, and stepper motors.

9.2 PROJECT 1—SIMPLE BRUSHED DC MOTOR CONTROL

9.2.1 Description

In this project a small BDC motor operating with 3–12 V is connected to one of the output ports of the nucleo-F411RE development board. The motor is turned ON for 10 s, then stopped for 5 s, and turned ON again for 10 s. This process is repeated continuously until stopped by the user. The motor speed is fixed in this project.

9.2.2 Aim

The aim of this project is to show how a small BDC motor can be connected and controlled from the Nucleo-F411RE development board.

9.2.3 Block Diagram

The BDC motors are low-cost, easy-to-drive electric motors that are used in many movement-based applications. These motors have two terminals and the voltage is applied across these terminals. The speed of the motor is directly proportional to the amount of the applied voltage. The direction of rotation can be changed easily by changing the polarity of the voltage across the terminals. The torque produced by the motor shaft is directly proportional to the current flowing through the motor windings. Fig. 9.1 shows the block diagram of

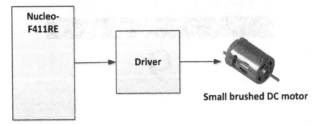

FIG. 9.1　Block diagram of the project.

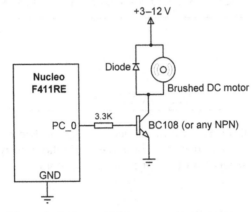

FIG. 9.2　Circuit diagram of the project.

the project. A driver circuit is used to power the motor as the current required by the motor is beyond the maximum capacity of a GPIO (general-purpose input/output) port.

9.2.4 Circuit Diagram

Fig. 9.2 shows the circuit diagram. The GPIO port PC_0 is connected to the motor through a switching transistor and a base resistor. The motor starts rotating when logic 1 is applied to the base of the transistor. Although a BC108-type transistor is used in this design, in general any NPN (negative-positive-negative)-type transistor can be used as long as the maximum current required by the motor is below the maximum allowable collector current of the transistor. Notice that a diode is connected across the motor terminals to protect the transistor from back emf generated by the motor windings.

9.2.5 The PDL

The algorithm of this project is very simple. Fig. 9.3 shows the operation of the program as a PDL.

9.2.6 Program Listing

The program listing (program: **SimpleMotor**) is shown in Fig. 9.4. At the beginning of the program GPIO port PC_0 is defined as an output and is assigned to variable Motor, ON and

```
BEGIN
    Configure PC_0 as digital output
    DO FOREVER
        Start the motor
        Wait 10 seconds
        Stop themotor
        Wait 5 seconds
    ENDDO
END
```

FIG. 9.3   Program PDL.

```
/********************************************************************

                SIMPLE DC MOTOR CONTROL
                =========================

In this project a small (3V to 12V) brushed DC motor is connected
to GPIO port PC_0. The motor is activated for 10 seconds, then
stopped for 5 seconds, and then activated fro 10 seconds. This
process is repeated until stopped manually.

Author: Dogan Ibrahim
Date   : September 2018
File   : SimpleMotor
********************************************************************/
#include "mbed.h"

DigitalOut Motor(PC_0);
#define ON 1
#define OFF 0

int main()
{
    Motor = 1;              // Motor ON
    wait(10.0);             // Wait 10 seconds
    Motor = 0;              // Motor OFF
    wait(5.0);              // Wait 5 seconds
}
```

FIG. 9.4   Program listing.

OFF are defined as 1 and 0, respectively. The program runs in an endless loop formed using a while statement. Inside this loop the motor is activated for 10 s, then stopped for 5 s, and then again activated for 10 s. This process continues until stopped manually by the user.

9.2.7 Using a Metal-Oxide-Semiconductor Field-Effect Transistor (MOSFET)

A MOSFET transistor can be used for motors requiring larger currents instead of a bipolar transistor. Fig. 9.5 shows the circuit diagram with an IRL540-type MOSFET transistor. This MOSFET has logic level gate drive and is therefore compatible with microcontroller output circuits. It is also possible to use a relay instead of transistor for even larger voltages and currents. This is shown in Fig. 9.6.

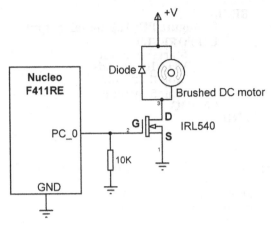

FIG. 9.5   Using a MOSFET Transistor.

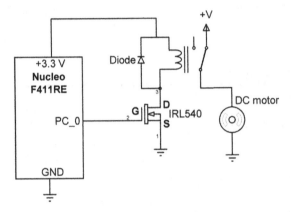

FIG. 9.6   Using a relay.

9.2.8 Suggestions for Additional Work

In this project the motor speed is the same when the motor is activated. Modify the program given in Fig. 9.4 by sending PWM waveform to the motor and change the speed of the motor by changing the Duty Cycle of this waveform. Try setting the Duty Cycle to 25%, 50%, and 100%.

9.3 PROJECT 2—CHANGING THE MOTOR ROTATION DIRECTION

9.3.1 Description

In this project the User button on the Nucleo-F411RE development board is used such that when the button is pressed the motor rotates in one direction, and when the button is released the motor rotates in the opposite direction. The motor speed is fixed in this project.

9.3.2 Aim

The aim of this project is to show how the rotation direction of the motor can easily be changed.

9.3.3 Block Diagram

The block diagram of the project is shown in Fig. 9.7. A direction control circuit (also known as H-bridge) is used to change the rotation direction of the motor when the button is pressed.

9.3.4 Circuit Diagram

Fig. 9.8 shows the circuit diagram of the project. In this project four MOSFET transistors are used to control the rotation direction of the motor. It is also possible to design this circuit using bipolar transistors. This circuit is also called the H-Bridge. The operation of the circuit is as follows:

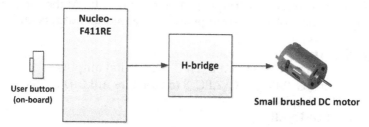

FIG. 9.7   Block diagram of the project.

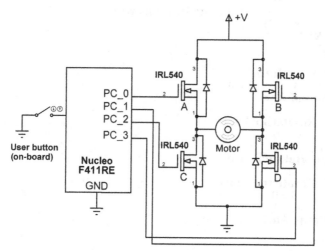

FIG. 9.8   Circuit diagram of the project.

A B C D	Motor Rotation
0 0 0 0	Motor stopped
1 0 0 1	Clockwise
0 1 1 0	Anticlockwise

The A, B, C, D terminals of the H-Bridge are connected to GPIO port pins PC_0, PC_1, PC_2, and PC_3 of the development board, respectively.

9.3.5 The PDL

The algorithm of this project is very simple. Fig. 9.9 shows the operation of the program as a PDL.

9.3.6 Program Listing

The program listing (program: **HBridge**) is shown in Fig. 9.10. At the beginning of the program, A, B, C, D terminals of the H-Bridge are configured as outputs and User button is

```
BEGIN/MAIN
        Configure PC_0,PC_1,PC_2,PC_3 as digital outputs
        Assign PC_0,PC_1,PC_2,PC_3 to variables A,B,C,D respectively
        CALL StopMotor
        DO FOREVER
            IF User button is pressed THEN
                CALL RotateClockwise
            ELSE
                CALL RotateAntiClockwise
            ENDIF
        ENDDO
END/MAIN

BEGIN/StopMotor
        A = B = C = D = 0
END/StopMotor

BEGIN/RotateClockwise
        B = C = 0
        A = D = 1
END/RotateClockwise

BEGIN/RotateAntiClockwise
        A = D = 0
        B = C = 1
END/RotateAntiClockwise
```

FIG. 9.9   Program PDL.

```
/*******************************************************************
                    DC MOTOR DIRECTION CONTROL
                    ===========================

In this project a small (3V to 12V) brushed DC motor is connected
to an H-Bridge made up of 4 MOSFET transistors. The rotation
direction of the motor is controlled by the User button on the
board. When the button is pressed the motor rotates clockwise,
otherwise it rotates anti-clockwise.

Author: Dogan Ibrahim
Date   : September 2018
File   : HBridge
*******************************************************************/
#include "mbed.h"

DigitalOut A(PC_0);
DigitalOut B(PC_1);
DigitalOut C(PC_2);
DigitalOut D(PC_3);
DigitalIn button(BUTTON1);

#define ON 1
#define OFF 0

//
// This function stops the motor
//
void StopMotor()
{
    A = OFF;
    B = OFF;
    C = OFF;
    D = OFF;
}


//
// This function rotates the motor clockwise
//
void RotateClockwise()
{
    B = OFF;
    C = OFF;
    A = ON;
    D = ON;
}


//
// This function rotates the motor anti-clockwise
//
void RotateAntiClockwise()
{
```

FIG. 9.10 Program listing.

(Continued)

```
            A = OFF;
            D = OFF;
            B = ON;
            C = ON;
    }


    //
    // Main Program. Check the button and control the direction
    //
    int main()
    {
        StopMotor();                   // Stop the motor at beginning

        if(button == 0)                // If button is pressed
            RotateClockwise();         // Rotate clockwise
        else                           // Button is not pressed
            RotateAntiClockwise();     // Rotate anti-clockwise
    }
```

FIG. 9.10, CONT'D

configured as input. Three functions are developed: **StopMotor, RotateClockwise**, and **RotateAntiClockwise**. At the beginning of the program the motor is stopped. Inside the main program the state of the User button is checked and if the button is pressed then the motor is rotated clockwise, otherwise it is rotated anticlockwise.

9.3.7 Suggestions for Additional Work

In this project the motor speed is the same when the motor is activated. Modify the program given in Fig. 9.10 by sending PWM waveform to the motor and change the speed of the motor by changing the Duty Cycle of this waveform. Try setting the Duty Cycle to 25%, 50%, and 100%.

9.4 PROJECT 3—SIMPLE SERVO MOTOR CONTROL

9.4.1 Description

In this project a small servo motor is connected to one of the GPIO ports of the Nucleo-F411RE development board. The servo motor is controlled as follows:

- position the servo moor all the way to the right (0 degrees)
- wait for 3 s
- position the servo motor all the way to the left (180 degrees)
- wait for 3 s
- position the servo motor in the middle (90 degrees)

9.4.2 Aim

The aim of this project is to show how a servo motor can be controlled.

9.4.3 Block Diagram

Servo motors are DC motors with built-in feedback circuits so that the positions of their shaft can be controlled accurately. These motors have three terminals: power, ground, and control. The control terminal is driven with PWM waveform usually with a period of 20 ms (50 Hz). In this project the small SG90 servo motor is used. Fig. 9.11 shows the shaft positions of this motor. When the PWM pulse duration is 1.5 ms then the shaft is in the middle position. This is also known as the 90 degrees position. When the PWM pulse duration is 2 ms then the shaft is all the way to the left, or 180 degrees. When the PWM pulse duration is 1 ms then the shaft is all the way to the right, or 0 degrees.

The SG90 has the following pin configuration:

Wire Color	Function
Brown	Ground
Red	+V
Orange	Control

The block diagram of the project is shown in Fig. 9.12. The SG90 servo motor has the following characteristics:

- Operating voltage: 3.3–6 V
- Running current: 220 ± 50 mA
- Weight: 9 g
- Speed: 01 s/60 degrees

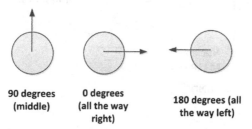

FIG. 9.11   Servo motor shaft positions.

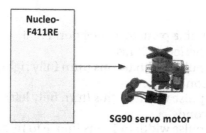

FIG. 9.12   Block diagram of the project.

9.4.4 Circuit Diagram

Fig. 9.13 shows the circuit diagram of the project. Notice that the servo motor is driven from the +5 V pin of the Nucle-F411RE development board (Pin 18, connector CN7). This pin has enough current capacity to drive the SG90 servo motor. In this project the servo motor control pin is connected to GPIO pin PA_8 which is a PWM channel pin.

9.4.5 The PDL

The algorithm of this project is very simple. Fig. 9.14 shows the operation of the program as a PDL.

9.4.6 Program Listing

The program listing (program: **Servo**) is shown in Fig. 9.15. At the beginning of the program variable **pwm** is assigned to PWM port PA_8. Inside the main program the period of the PWM waveform is set to 20 ms. The servo motor shaft is then turned fully right by setting the pulse width to 1 ms. After 3 s of delay the pulse width is set to 2 ms which then turns the motor shaft fully left. Finally, the motor shaft is moved to the middle position.

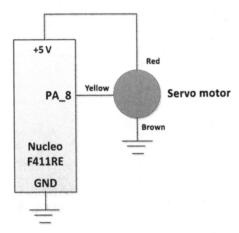

FIG. 9.13   Circuit diagram of the project.

```
BEGIN
    Assign variable pwm to analog port PA_8
    Set PWM period to 20ms
    Set PWM pulse width to 1ms (turn fully right)
    Wait 3 seconds
    Set PWM pulse width to 2ms (turn fully left)
    Wait 3 seconds
    Set PWM pulse width to 1.5ms (move to middle position)
END
```

FIG. 9.14   Program PDL.

```
/***********************************************************************

                        SERVO MOTOR CONTROL
                        ===================

In this project a small servo motor (SG90) is connected to PWM port PA_8
of the Nucleo-F411RE development board. The program turns the motor fully
to the right, waits 3 seconds, then fully to the left, waits for 3 seconds,
and then moves it to the middle.

Author: Dogan Ibrahim
Date  : September 2018
File  : Servo
***********************************************************************/
#include "mbed.h"

PwmOut pwm(PA_8);                          // Analog port
float period = 0.02;                       // Period (20ms)
float left, right, middle;

int main()
{
    left = 0.002;                          // 2ms
    middle = 0.0015;                       // 1.5ms
    right = 0.001;                         // 1ms

    pwm.period(period);                    // Set the period
    pwm.pulsewidth(right);                 // To the right
    wait(3.0);                             // Wait 3 seconds
    pwm.pulsewidth(left);                  // To the left
    wait(3.0);                             // Wait 3 seconds
    pwm.pulsewidth(middle);                // To the middle
}
```

FIG. 9.15   Program PDL.

9.4.7 Suggestions for Additional Work

Modify this project by connecting a potentiometer to one of the analog input ports of the development board. Then, write a program to turn the servo motor shaft as the potentiometer arm is rotated left and right.

9.5 PROJECT 4—SIMPLE STEPPER MOTOR CONTROL

9.5.1 Description

In this project a small stepper motor is connected to the Nucleo-F411RE development board through a driver module. The motor shaft is rotated as follows:

- three complete revolutions in one direction
- stop for 3 s
- three complete revolutions in the reverse direction

9.5.2 Aim

The aim of this project is to show how a stepper motor can be controlled.

9.5.3 Block Diagram

Stepper motors rotate by small steps in response to applied voltage pulses. The motors have several windings and the voltage steps must be applied in the correct sequences to turn the motor shaft. Basically, there are two types of stepper motors: unipolar and bipolar.

Unipolar Stepper Motors

As shown in Fig. 9.16, these motors have four windings and five or six leads depending on the whether or not the common leads are joined together. Unipolar motors can be operated in full-step or half-step modes. Half-step mode requires more steps for rotation but provides higher precision. Full-step modes can be either one phase or two phase. In one-phase mode only one winding is excited at any time, while in two-phase mode two windings are excited at the same time. Two-phase mode gives higher torque. Table 9.1 illustrates the two-phase full-step mode of operation which is the mode used in this project.

Bipolar Stepper Motors

Bipolar stepper motors have two windings as shown in Fig. 9.17. The drive sequence of these motors is presented in Table 9.2.

The block diagram of this project is shown in Fig. 9.18. A small 28BYJ-48-type unipolar geared stepper motor is used in this project. This motor has the following features:

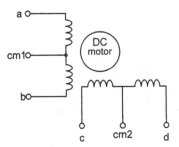

FIG. 9.16   Unipolar stepper motor windings.

TABLE 9.1   2 Phase Full-Step Mode of Excitation

Step	a	c	b	d
1	1	0	0	1
2	1	1	0	0
3	0	1	1	0
4	0	0	1	1

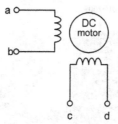

FIG. 9.17   Bipolar stepper motor windings.

TABLE 9.2   Bipolar Motor Excitation

Step	a	c	b	d
1	+	−	−	−
2	−	+	−	−
3	−	−	+	−
4	−	−	−	+

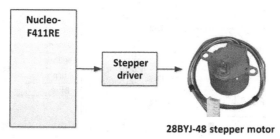

28BYJ-48 stepper motor

FIG. 9.18   Block diagram of the project.

- Operating voltage: 5 V
- Number of phases: 4
- Gear ratio: 64

The common pins of the windings are joined together and as a result the motor has only five terminals. The terminals are identified as follows:

Pin	Description
2 (Pink)	Winding 1 terminal
4 (Orange)	Winding 1 terminal
5 (Red)	Winding1 + Winding 2 common terminal
3 (Yellow)	Winding 2 terminal
1 (Blue)	Winding 2 terminal

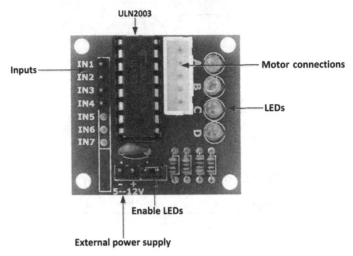

FIG. 9.19   ULN2003 stepper driver module.

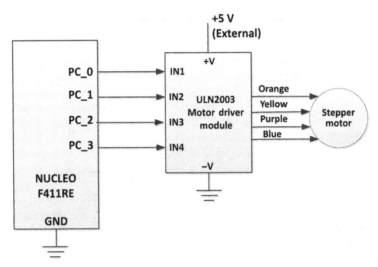

FIG. 9.20   Circuit diagram of the project.

9.5.4 Circuit Diagram

In this project a ULN2003-type stepper motor driver module is used. This module (see Fig. 9.19) has four inputs IN1, IN2, IN3, and IN4. Additionally four LEDs are provided named as A, B, C, and D. The LEDs indicate the state of the four inputs. These LEDs can be enabled or disabled by a short jumper provided at the bottom part of the module. External power (+5 to +12V) for the motor is applied through the connectors provided at the bottom part of the module (next to the LED enable/disable jumper). The motor is connected to the four connectors located at the top part of the module next to the LEDs.

The circuit diagram of the project is shown in Fig. 9.20. GPOI pins PC_0, PC_1, PC_2, and PC_3 of the Nucleo-F411RE development board are connected to pins IN1, IN2, IN3, and IN4

BEGINMAIN
  Define connections A,B,C,D to PC_0 to PC_3
  Define motor parameters
  Store steps in array FullMode
  CALL Clockwise(3) to make 3 complete clockwise revolutions
  Wait 3 seconds
  CALL AntiClockwise(3) to make 3 complete anti-clockwise revolutions
  Wait 3 seconds
END/MAIN

BEGIN/Clockwise(n)
  DO n times
      DO for one cycle time
          CALL Pulse to send pulses to the motor in clockwise mode
      ENDDO
  ENDDO
END/Clockwise

BEGIN/AntiClockwise(n)
  DO n times
      DO for k from 0 to 4
          CALL Pulse(k) to send pulses to the motor in anti-clockwise mode
      ENDDO
  ENDDO
END/AntiClockwise

BEGIN/Pulse(k)
  CALL CheckBit(k,0) to extract bit 0 of index k of FullMode
  Send bit to A
  CALL CheckBit(k,1) to extract bit 1 of index k of FullMode
  Send bit to B
  CALL CheckBit(k,2) to extract bit 2 of index k of FullMode
  Send bit to C
  CALL CheckBit(k,3) to extract bit 3 of index k of FullMode
  Send bit to D
END/Pulse

BEGIN/CheckBit(p, m)
  Extract bit m of array FullMode indexed by p
END/CheckBit

FIG. 9.21 Program PDL.

of the ULN2003 module, respectively. A 5 V supply is applied from the development board pin 18 of connector CN7. The stepper motor is connected to the ULN2003 module as shown in the figure.

9.5.5 The PDL

The PDL of the program is shown in Fig. 9.21.

9.5.6 Program Listing

In this project the stepper moor is operated in two-phase full-step mode. The 28BYJ-48 stepper motor has four steps per cycle and 11.25 degrees/step ($4 \times 11.25 = 45$ degrees/cycle).

TABLE 9.3   Step Sequence for One Cycle

Step	4 (Orange) IN1	3 (Yellow) IN2	2 (Pink) IN3	1 (Blue) IN4
1	1	1	0	0
2	0	1	1	0
3	0	0	1	1
4	1	0	0	1

This corresponds to $360/11.25 = 32$ steps for one complete revolution of the *internal motor shaft*. The motor is geared with a gear ratio of 64 (it is exactly 63.68395), the number of steps for one complete revolution of the *external motor shaft* is $32 \times 64 = 2048$ steps/revolution, which corresponds to $2048/4 = 512$ cycles with 4 steps/cycle. Therefore, we have to send 512 cycles $\times$ 4 pulses/cycle $= 2048$ steps to the motor for one complete revolution of its external shaft. Table 9.3 illustrates the step sequence for one cycle.

The speed of a stepper motor depends on the delay inserted between each step. In full-step mode there are 2048 steps in a complete external revolution and therefore the speed of the motor in RPM (revolutions per minute) is given by

$$RPM = 60 \times 10^3 / (2048 \times T)$$

or

$$RPM = 29.3/T$$

where RPM is the motor speed, and T is the delay between each step in milliseconds. The above equation can be written as

$$T = 0.0293/RPM$$

where T is in seconds. In this program we will choose the motor speed as 12 RPM, which gives the delay between each step as $T = 0.0293/12 = 0.00244$ s.

The program listing (program: **Stepper**) is shown in Fig. 9.22. At the beginning of the program the connection between the development board and the driver module are defined and various motor parameters such as the **RPM, CyclesPerRev**, and **StepsPerCycle** are defined. Notice that the speed of the motor is set to 12 RPM. The pulses to be sent in a cycle in full-step mode are stored in array **FullMode**. The program consists of a number of functions. Function **Clockwise** receives the argument **RevCount** and sends pulses to the motor to make complete **RevCount** clockwise rotations. Similarly, function **AntiClockwise** makes complete **RevCount** anticlockwise rotations. Function **Pulse** sends a cycle of pulses (four pulses) to the motor by calling function **CheckBit** which extracts the bits of array **FullMode** for a given step and sends the extracted bits to the motor windings. The main program calculates the delay between each step in seconds, calls function **Clockwise** with argument set to 3 to make three complete revolutions. After 3 s of delay function **AntiClockwise** is called with argument 3 so that three complete anticlockwise revolutions are made.

```
/************************************************************************

                        STEPPER MOTOR CONTROL
                        ====================

In this project a small unipolar stepper motor is connected to the Nucleo-
F411 development board through a driver module. The motor rotates 3 times
in clockwise direction, then stops for 3 seconds, and then rotates 3 timee
in anti-clockwise direction.

Author: Dogan Ibrahim
Date   : September 2018
File   : Stepper
************************************************************************/
#include "mbed.h"

//
// GPIO connections to ULN2003 module
//
DigitalOut A(PC_0);                          // Connection to IN1
DigitalOut B(PC_1);                          // Conenction to IN2
DigitalOut C(PC_2);                          // Connection to IN3
DigitalOut D(PC_3);                          // Connection to IN4

int RPM = 12;                                // RPM = 12
int CyclesPerRev = 512;                      // Cycles per rev
int StepsPerCycle = 4;                       // Steps per cycle
float StepDelay;                             // Step delay
int FullMode[4] = {0x0C, 0x06, 0x03, 0x09};

//
// This function checks the bits of array Full-Mode and returns 0 or 1
// depending whether or not the bit position is 0 or 1 respectively
//
int CheckBit(int p, int m)
{
    int y;

    y = 1 << m;
    if((FullMode[p] & y) != 0)
          return 1;
    else
          return 0;
}


//
// This function sends pulses (0 or 1) in full-mode
//
void Pulse(int k)
{
    A = CheckBit(k, 0);
    B = CheckBit(k, 1);
    C = CheckBit(k, 2);
    D = CheckBit(k, 3);
}
```

FIG. 9.22   Program listing.

(Continued)

```
//
// This function rotates the motor shaft clockwise
//
void Clockwise(int RevCount)
{
    for(int j = 0; j < RevCount; j++)
    {
        for(int m = 0; m < CyclesPerRev; m++)
        {
            for(int i = StepsPerCycle; i >= 0; i--)
            {
                Pulse(i);
                wait(StepDelay);
            }
        }
    }
}


//
// This function rotates the motor shaft anti-clockwise
//
void AntiClockwise(int RevCount)
{
    for(int j = 0; j < RevCount; j++)
    {
        for(int m = 0; m < CyclesPerRev; m++)
        {
            for(int i = 0; i < StepsPerCycle; i++)
            {
                Pulse(i);
                wait(StepDelay);
            }
        }
    }
}


//
// Main Program
//
int main()
{
    StepDelay = 0.0293 / RPM;        // In seconds
    Clockwise(3);                    // 3 revolutions
    wait(3.0);                       // Wait 3 seconds
    AntiClockwise(3);                // 3 revolutions
    wait(3.0);                       // Wait 3 seconds
}
```

FIG. 9.22, CONT'D

9.6 SUMMARY

In this chapter we have learned about the following:

- Types of electric motors
- Controlling BDC motors
- Controlling servo motors
- Types of stepper motors
- Controlling unipolar stepper motors

9.7 EXERCISES

1. Explain how the direction of rotation of a BDC motor can be changed.
2. Explain the differences between a BDC motor and a stepper motor.
3. Draw the circuit diagram to show how a bipolar stepper motor can be connected to a Nucleo-F411RE development board.
4. A unipolar stepper motor with a stepping angle of 10 degrees is connected to a Nucleo-F411RE development board. Write a program to rotate the motor by 90 degrees.
5. Explain the differences between a unipolar and a bipolar stepper motor.

Using Liquid Crystal Displays (LCDs)

10.1 OVERVIEW

In this chapter we shall be developing projects using LCDs with the Nucleo-F411RE development board. The highly popular HD44780 (or compatible)-type LCD is used in all the projects in this chapter.

In microcontroller-based systems we usually want to interact with the system, for example, to enter a parameter, to change the value of a parameter, or to display the output of a measured variable. Data is usually entered to a system using a switch, a small keypad, or a fully blown keyboard. Data is usually displayed using an indicator such as one or more LEDs, 7-segment displays, or LCD-type displays. The LCDs have the advantages that they can display alphanumeric as well as graphical data. Some LCDs have 40 or more character lengths with the capability to display data in several lines. Some other LCDs can be used to display graphical images (graphical LCDs, or simply GLCDs), such as animation. Some displays are in single or in multicolor, while some others incorporate back lighting so that they can be viewed in dimly lit conditions.

The LCDs can be connected to a microcontroller either in parallel form or in serial form. Parallel LCDs (e.g., Hitachi HD44780) are connected using more than one data line and several control lines and the data is transferred in parallel form. It is common to use either four or eight data lines and two or more control lines. Using a four-wire connection saves I/O (input/output) pins but it is slower since the data is transferred in two stages. Serial LCDs on the other hand are connected to a microcontroller using only one data line. In these type of LCDs data is usually sent to the LCD using the standard RS-232 asynchronous data communication protocol or the I^2C bus protocol. Serial LCDs are in general much easier to use and require less wiring, but they cost more than the parallel ones.

The programming of LCDs is a complex task and requires a good understanding of the internal operations of the LCD controllers, including knowledge of their exact timing requirements. Fortunately, most high-level languages, including Mbed, provide special library functions to control LCDs as well as GLCDs for displaying alphanumeric or graphical data. All the user has to do is to connect the display to the microcontroller, define the connections between the microcontroller and the display device in software, and then send commands to display the required data on the LCD or the GLCD.

10.1.1 HD44780 LCD Module

Although there are several types of LCDs, the HD44780 is currently one of the most popular LCD modules used in industry and also by hobbyists. This module is an alphanumeric monochrome display and comes in different sizes. Modules with 16 columns are popular in most small applications, but other modules with 8, 20, 24, 32, and 40 columns are also available. Although most LCDs have two lines (or rows) as the standard, it is possible to purchase models with one or four lines. The LCD displays are available with standard 14-pin connectors, although 16-pin modules are also available, providing terminals for backlighting. Table 10.1 illustrates the pin configuration and pin functions of a 14-pin LCD module. A brief summary of the pin functions is given below:

V_{SS} (pin 1) and V_{DD} (pin 2) are the ground and power supply pins. Although the manufacturers specify 5 V DC supply, the modules will usually work with as low as 3.6 V or as high as 6 V.

V_{EE} is pin 3 and this is the contrast control pin used to adjust the contrast of the display. The arm of a 10 K potentiometer is normally connected to this pin and the other two terminals of the potentiometer are connected to the ground and power supply pins. The contrast of the display is adjusted by rotating the potentiometer arm.

Pin 4 is the Register Select (RS) and when this pin is LOW, data transferred to the display is treated as commands. When RS is HIGH, character data can be transferred to and from the display.

Pin 5 is the Read/Write (R/W) line. This pin is pulled LOW in order to write commands or character data to the LCD module. When this pin is HIGH, character data or status information can be read from the module. This pin is normally connected permanently LOW so that commands or character data can be sent to the LCD module.

TABLE 10.1 Pin Configuration of HD44780 LCD Module

Pin No	Name	Function
1	V_{SS}	Ground
2	V_{DD}	+ve supply
3	V_{EE}	Contrast
4	RS	Register select
5	R/W	Read/write
6	E	Enable
7	D0	Data bit 0
8	D1	Data bit 1
9	D2	Data bit 2
10	D3	Data bit 3
11	D4	Data bit 4
12	D5	Data bit 5
13	D6	Data bit 6
14	D7	Data bit 7

Enable (E) is pin 6 which is used to initiate the transfer of commands or data between the LCD module and the microcontroller. When writing to the display, data is transferred only on the HIGH to LOW transition of this pin. When reading from the display, data becomes available after the LOW to HIGH transition of the enable pin and this data remains valid as long as the enable pin is at logic HIGH.

Pins 7–14 are the eight data bus lines (D0–D7). Data can be transferred between the microcontroller and the LCD module using either a single 8-bit byte, or as two 4-bit nibbles. In the latter case only the upper four data lines (D4–D7) are used. The 4-bit mode has the advantage that four less I/O lines are required to communicate with the LCD. The 4-bit mode is however slower since the data is transferred in two stages. In this book we shall be using the 4-bit interface only.

In 4-bit mode the following pins of the LCD are used. The R/W line is permanently connected to ground. This mode uses six GPIO port pins of the microcontroller:

$$V_{SS}, V_{DD}, V_{EE}, E, R/S, D4, D5, D6, D7$$

10.2 PROJECT 1—DISPLAYING TEXT ON THE LCD

10.2.1 Description

In this project a 2 line by 16 character LCD is connected to the Nucleo-F411RE development board. The text **Nucleo-F411RE** is displayed on the LCD.

10.2.2 Aim

The aim of this project is to show how text can be displayed on an LCD using Mbed with the Nuclo-F411RE development board.

10.2.3 Block Diagram

The block diagram of the project is shown in Fig. 10.1.

10.2.4 Circuit Diagram

Fig. 10.2 shows the circuit diagram of the project. Since the LCD is operated in 4-bit mode, only six GPIO pins are used as mentioned earlier to interface with the LCD. The connections between the LCD and the Nucleo-F411RE development board are as follows (GPIO connections are shown in bold):

LCD Pin	Nucleo-F411RE Pin	Connector	Description
V_{SS}	GND	Pin 8, CN7	Ground
V_{DD}	+5V	Pin 18, CN7	Power

Continued

LCD Pin	Nucleo-F411RE Pin	Connector	Description
V_{EE}			Potentiometer arm
R/W	GND	Pin 8, CN7	Ground
R/S	PC_0	Pin 38, CN7	Digital output
E	PC_1	Pin 36, CN7	Digital output
D4	PC_2	Pin 35, CN7	Digital output
D5	PC_3	Pin 37, CN7	Digital output
D6	PC_4	Pin 34, CN10	Digital output
D7	PC_5	Pin 6, CN10	Digital output

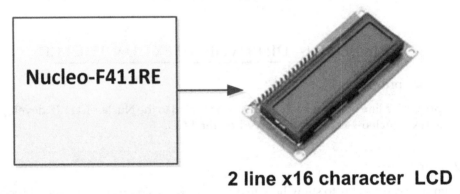

2 line x16 character LCD

FIG. 10.1   Block diagram of the project.

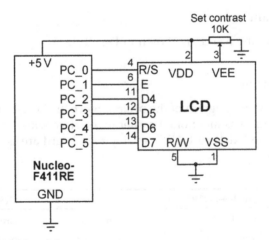

FIG. 10.2   Circuit diagram of the project.

10.2.5 Program Listing

There are several LCD libraries for Mbed. In this project we have used the one called **TextLCD**. Before using the library we must import it into our workspace. The steps are as follows:

- Right click on your project folder (**LCDText**) in the **Program Workspace**.
- Click **Import Library** -> **From Import Wizard**....
- Enter **TextLCD** in the search box at the top right-hand side of the screen (see Fig. 10.3) and click **Search**.
- Double click on **TextLCD** with the tag **HD44780 TextLCD**.
- You should see the library listed in your project folder.

The program listing (program: **LCDText**) is shown in Fig. 10.4. At the beginning of the program header files **mbed.h** and **TextLCD.h** are included in the program. The interface between the LCD and the development board must be defined at the beginning of the program by specifying the GPIO pins that are connected to the LCD pins in the following order:

R/S, E, D4, D5, D6, D7

Inside the main program the text Nucleo-F411RE is displayed on the LCD.

10.2.6 LCD Library Functions

The LCD library used in this project supports a number if useful functions for controlling the text to be displayed on the LCD. These functions are as follows:

cls: Clear the screen
putc: Write a character on the LCD
printf: Write a formatted string on the LCD
locate(column, line): Position the cursor at the given column and line

10.2.7 Suggestions for Additional Work

Modify the program given in Fig. 10.4 so that the text is displayed starting from column 2 of the second line of the LCD.

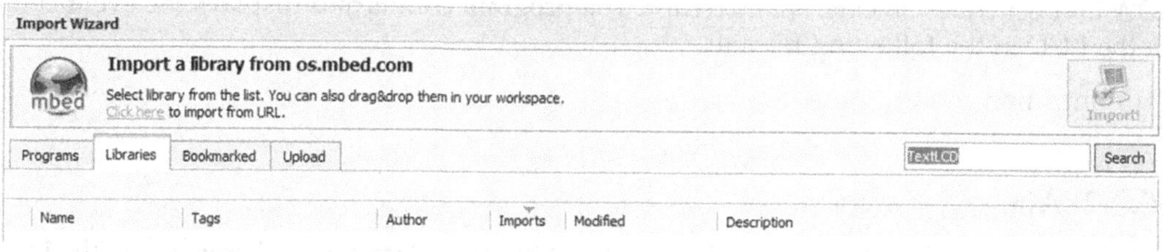

FIG. 10.3   Search for the LCD library.

```
/**********************************************************************

                            LCD TEXT
                            ========

In this project a HD44780 controller type LCD is connected to PORTC
of the Nucleo-F411RE development board as follows:

R/S PC_0
E   PC_1
D4  PC_2
D5  PC_3
D6  PC_4
D7  PC_5

The program displays the text "Nucleo-F411RE" on the LCD

Author: Dogan Ibrahim
Date  : September 2018
File  : LCDText
**********************************************************************/
#include "mbed.h"
#include "TextLCD.h"


TextLCD MyLCD(PC_0, PC_1, PC_2, PC_3, PC_4, PC_5);        // LCD interface

int main()
{
    MyLCD.printf("Nucleo-F411RE");                        // Display text
}
```

FIG. 10.4   Program listing.

10.3 PROJECT 2—EXTERNAL INTERRUPT-DRIVEN EVENT COUNTER

10.3.1 Description

In this project a character LCD is connected to the development board as in the previous project. A GPIO pin is configured as an external interrupt pin and this pin is used to simulate the occurrence of external events. An event is said to occur when this pin goes from HIGH to LOW and generates an external interrupt. The program counts and displays the event count on the LCD in the following format:

Count=nnn

10.3.2 Aim

The aim of this project is to show how text and numeric data can be displayed on the LCD. Additionally, the project shows how an external interrupt pin can be used in a project.

10.3.3 Block Diagram

The block diagram of the project is shown in Fig. 10.5.

10.3.4 Circuit Diagram

Fig. 10.6 shows the circuit diagram of the project. A push-button switch is connected to GPIO pin PA_0 of the development board such that the state of the button is normally at logic HIGH and it goes to logic LOW when the button is pressed. The LCD is connected to PORTC of the development board as in the previous project.

10.3.5 The PDL

The program PDL is shown in Fig. 10.7.

10.3.6 Program Listing

Mbed provides several functions for handling external interrupts. Table 10.2 presents a list of the external interrupt functions. In this project we will be generating interrupts when GPIO pin PA_0 edge falls from HIGH to LOW.

Fig. 10.8 shows the program listing (program: **ExtInt**). At the beginning of the program **mbed.h** and the LCD header file **TextLCD.h** (see previous project. You can copy and paste

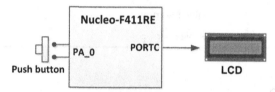

FIG. 10.5   Block diagram of the project.

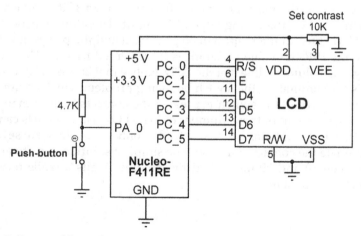

FIG. 10.6   Circuit diagram of the project.

BEGIN/MAIN
　　Define LCD to development board interface
　　Initialize Count to 0
　　Attach function ISR to falling edge of external interrupts on PA_0
　　DO FOREVER
　　　　Display current Count
　　　　Wait for switch debounce
　　ENDDO
END/MAIN

BEGIN/ISR
　　Increment Count by one
END/ISR

FIG. 10.7   Program PDL.

TABLE 10.2   Mbed External Interrupt Functions

Interrupt Function	Description
InterruptIn	Create an external interrupt on specified pin
rise	Attach to interrupt service routine when the input edge rises
fall	Attach to interrupt service routine when the input edge falls
enable_irq	Enable interrupts
disable_irq	Disable interrupts

folder **TextLCD** from the previous project folder to this project folder) are included in the program. GPIO pin PA_0 is configured as an external interrupt pin, variable **Count** is initialized to 0, and the interface between the LCD and the Nucleo-F411RE development board is defined. Inside the main program a function called **ISR** is attached to external interrupts on pin PA_0 on the falling edge of its input. Therefore, whenever PA_0 input goes from HIGH to LOW an interrupt will be generated and the program will jump to this function. Inside the **ISR** the value of **Count** is incremented by one. The current value of *Count* is displayed at column 0, line 0 of the LCD. Notice that 20-ms delay is inserted inside the loop. This is for eliminating the switch bouncing problem which commonly occurs in mechanical switches. When the switch is pressed the contacts bounce many times before they settle down. As a result of this a number of logic LOWs and HIGHs can be generated which can increase the Count many times instead of only once. There are several hardware solutions to eliminate the contact bouncing. The simplest solution in software is to insert a small delay between 10 and 20 ms so that the program waits a while until the contacts settle down to their final state.

```
/*********************************************************************
                EXTERNAL INTERRUPT BASED EVENT COUNTER
                ========================================

In this project a HD44780 controller type LCD is connected to PORTC
of the Nucleo-F411RE development board. Additionally a push-button
switch is conencted to GPIO pin PA_0. External interrupts are
configured in falling mode on PA_0 so that the program jumps to the
interrupt service routine whenever PA_0 edge goes from HIGH to LOW.
A counter is incremented when an interrupt occurs and the count is
displayed on the LCD.

Author: Dogan Ibrahim
Date  : September 2018
File  : ExtInt
*********************************************************************/
#include "mbed.h"
#include "TextLCD.h"

InterruptIn button(PA_0);

int Count = 0;

TextLCD lcd(PC_0, PC_1, PC_2, PC_3, PC_4, PC_5);        // LCD interface

//
// External interrupt service routine. The program
// jumps here when an external interrupt is generated
//
void ISR()
{
    Count++;                                            // Increment Count
}


//
// Main program
//
int main()
{
    button.fall(&ISR);                          // Attach external interrupt
    lcd.cls();                                  // Clear LCD screen

    while(1)                                    // Do forever
    {
        lcd.locate(0, 0);                       // Position (0,0) of LCD
        lcd.printf("Count = %d", Count);        // Display Count
        wait_ms(20);                            // Wait 20ms
    }
}
```

FIG. 10.8   Program listing.

10.4 PROJECT 3—ULTRASONIC HEIGHT MEASUREMENT

10.4.1 Description

Having the correct height is very important especially during the child development ages. Human height is usually measured using a stadiometer. A stadiometer can be either mechanical or electronic. Most stadiometers are portable, although they can also be wall mounted. A mechanical stadiometer is commonly used in schools, clinics, hospitals, or doctors' offices. This project is about designing a microcontroller-based device to measure the human height using ultrasonic techniques.

As shown in Fig. 10.9 a mechanical stadiometer consists of a long ruler, preferably mounted vertically on a wall, with a movable horizontal piece that rests on the head of the person whose height is being measured. The height is then read on the ruler corresponding to the point of the horizontal piece. Using such devices the height can be measured from about 14–200 cm with graduations of 0.1 cm.

In this project we will see how to design of an electronic stadiometer based on the principle of ultrasonic waves.

10.4.2 Aim

The aim of this project is to show how an ultrasonic height measurement device can be designed using Mbed and the Nucleo-F411RE development board.

FIG. 10.9   Mechanical stadiometer.

FIG. 10.10 Block diagram of the height measurement system.

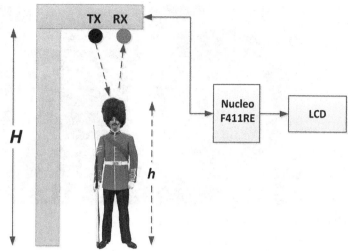

10.4.3 Block Diagram

Fig. 10.10 shows the block diagram of the height measurement system. Basically, a pair of ultrasonic transducers (a transmitter TX, and a receiver RX) are mounted at the top of a pole whose exact height from the ground level is known, say **H**. The person whose height is to be measured stands under the ultrasonic transducers. The system sends an ultrasonic signal through the TX module. This signal hits the person's head and is received by the RX module. By knowing the speed of sound in the air and the time the signal takes to return, we can calculate the distance from the transducers to the head of the person. If this distance is called **h**, then the height of the person is simply given by the difference **H − h**.

10.4.4 Circuit Diagram

In this project a HC-SR04-type ultrasonic transmitter/receiver module is used (see Fig. 10.11). This module has the following specifications:

- Operating voltage: 5 V
- Operating current: 2 mA
- Detection distance: 2–450 cm
- Input trigger signal: 10 μs TTL
- Sensor angle: not more than 15 degrees

HC-SR04 is a 4-pin device with the following pin names and descriptions:

Vcc: Power input
Trig: Trigger input
Echo: Echo output
Gnd: Power ground

FIG. 10.11   HC-SR04 ultrasonic module.

The basic principle of operation of the HC-SR04 ultrasonic sensor module is as follows (see Fig. 10.12):

- A 10 μs trigger pulse is sent to the module
- The module then sends eight 40 kHz square wave signals to the target and sets the echo pin HIGH
- The program starts a timer
- The signal hits the target and echoes back to the module
- When the signal is returned to the module the echo pin goes LOW
- The timer is stopped
- The duration of the echo signal is calculated and this is proportional to the distance to the target

The distance to the object is calculated as follows:

$$\text{Distance to object (in meters)} = (\text{duration of echo time in seconds}^* \text{speed of sound})/2$$

The speed of sound is 340 m/s, or 0.034 cm/μs.
Therefore,

$$\text{Distance to object (in cm)} = (\text{duration of echo time in μs})^* \, 0.034/2$$

or

$$\text{Distance to object (in cm)} = (\text{duration of echo time in μs})^* \, 0.017$$

For example, if the duration of the echo signal is 294 μs then the distance to the object is calculated as follows:

$$\text{Distance to object (cm)} = 294^* \, 0.017 = 5 \text{cm}$$

Fig. 10.13 shows the circuit diagram of the project. An LCD, is connected to PORTC of the development board as in the previous project. The **trig** and **echo** pins of the HC-SR04 sensor module are connected to GPIO pins PC_6 and PC_7, respectively.

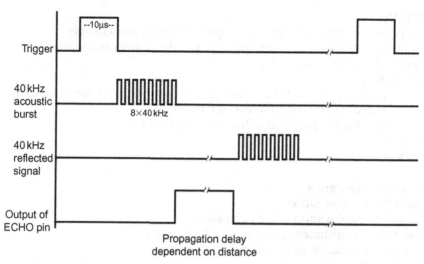

FIG. 10.12 Operation of the ultrasonic sensor module.

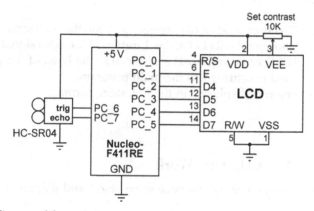

FIG. 10.13 Circuit diagram of the project.

10.4.5 The PDL

Fig. 10.14 shows the program PDL.

10.4.6 Program Listing

The program listing (program: **Height**) is shown in Fig. 10.15. At the beginning of the program variables **trig** and **echo** are assigned to GPIO ports PC_6 and PC_7, respectively, and the interface between the LCD and the development board are defined. The height of the sensor (variable **H**) is set to 2 m (200 cm). The main program calls function **CalculateDistance** to calculate the distance between the sensor and the head of the person. This value is then subtracted from the sensor height to find the height of the person which is then displayed at column 0, line 0 of the LCD. The program repeats after 1 s delay.

BEGIN/MAIN
> Configure trig and echo pins as output and input respectively
> Define the interface between the LCD and the development board
> Set height of the sensor (H)
> **DO FOREVER**
>> CALL CalculateDistance to find the distance (h) to the head of the person
>> Calculate the height of the person (H-h)
>> Display the height at column 0, line 0 of LCD
> **ENDDO**
END/MAIN

BEGIN/CalculateDistance
> Send 10μs trigger pulse
> Calculate the duration of the echo signal
> Calculate the distance to the head of the person
> Return the distance to the main program
END/CalculateDistance

FIG. 10.14   Program PDL.

Function **CalculateDistance** sends a 10 μs trigger pulse to the sensor module and resets the Timer to 0. When the echo signal falls LOW the duration of the echo signal is read and stored in floating point variable **EchoDuration**. The distance to the head of the person is then calculated in centimeters and is returned to the main program.

The height of the person is displayed in the following format:

nnn.nn cm

10.4.7 Suggestions for Additional Work

Modify the project to measure the distance to an object and display it on an LCD.

10.5 PROJECT 4—CALCULATOR USING A KEYPAD

10.5.1 Description

In this project a basic four-function integer calculator is designed using a 4×4 keypad. The calculator can perform addition, subtraction, multiplication, and division. The calculator operates in Reverse Polish Notation (RPN) where the first number is entered, followed by the Enter key, the second number is entered again followed by the Enter key, then the required operation is entered. The result is displayed on the LCD.

The heading CALCULATOR is displayed at column 0, row 0 of the LCD for 2 s. As an example, the following steps are required to add numbers 20 and 10 (characters typed by the user are in bold for clarity). The result is displayed for 5 s and then the LCD is cleared, ready for the next calculation:

```
/***********************************************************************

                   ULTRASONIC HEIGHT MEASUREMENT
                   ==============================

In this project a HD44780 controller type LCD is connected to PORTC of
the Nucleo-F411RE development board. Additionally, an ultrasonic TX,RX
pair is used in the project. The project calculates the height of a
person and displays it on the LCD. The height of the sensor is set to
2m (200cm). The height of the person is calculated as H-h where H is
the height of the sensor and h is the distance from the sensor to the
head of the person

Author: Dogan Ibrahim
Date  : September 2018
File  : Height
***********************************************************************/
#include "mbed.h"
#include "TextLCD.h"

DigitalOut trig(PC_6);                              // trig output
DigitalIn echo(PC_7);                               // echo input
Timer tmr;                                          // Timer

TextLCD lcd(PC_0, PC_1, PC_2, PC_3, PC_4, PC_5);    // LCD interface

float H = 200.0;                                    // Height of sensor

//
// This function calculates the distance to the head of the person
// and returns the result to the main program in centimetres
//
float CalculateDistance()
{
    float EchoDuration, DistToHead;

    trig = 0;
    wait_ms(80);                        // Wait to settle

    trig = 1;                           // Set trigger HIGH
    wait_us(10);                        // 10us trigger pulse
    trig = 0;                           // Set triger low

    while(!echo);                       // Wait for echo

    tmr.reset();                        // Reset Timer to 0
    while(echo);                        // Wait if echo is HIGH

    EchoDuration = tmr.read_us();       // Read echo duration
    DistToHead = EchoDuration * 0.017f; // Distance to head (cm)
    return DistToHead;                  // Return distance
}
```

FIG. 10.15 Program listing.

(Continued)

```
//
// Main program. Call CalculateDistance to find teh distance from the
// sensor to the head of the person. Then calculate the height of the
// person by subtracting this value from the height of the sensor
// which is set to 2m (200cm)
//
int main()
{
    float Distance, Height;

    lcd.cls();                              // Clear LCD screen
    tmr.start();                            // Start Timer

    while(1)                                // Do Forever
    {
        Distance = CalculateDistance();     // Diatance to head
        Height = H - Distance;              // Calculate height
        lcd.locate(0, 0);                   // LCD to 0,0
        lcd.printf("%6.2f cm  ", Height);   // Display height
        wait(1.0);                          // Wait 1 second
    }
}
```

FIG. 10.15,   CONT'D

No1: **20** <Enter>
No2: **10** <Enter>
Op: +
Res = 30

10.5.2 Aim

The aim of this project is to show how a keypad can be used with Mbed and the Nucleo-F411RE development board in a project.

10.5.3 Block Diagram

Keypads are commonly used in microcontroller-based applications to enter numeric or alphanumeric data to the microcontroller. The most widely used type is a 4×4 keypad consisting of 16 buttons, arranged in the form of four columns and four rows. The keypad used in this project is the 4×4 keypad manufactured by mikroElektronika (www.mikroe.com) and is shown in Fig. 10.16.

The block diagram of the project is shown in Fig. 10.17.

10.5.4 Circuit Diagram

The mikroElektronika 4×4 keypad is interfaced to a microcontroller using a 10-way IDC (insulation-displacement contact) connector. Fig. 10.18 shows the connection diagram of the keypad. The row signals are named as P4, P5, P5, and P7. Similarly, the column signals are

FIG. 10.16   4 × 4 keypad.

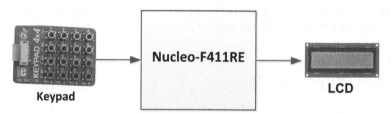

FIG. 10.17   Block diagram of the project.

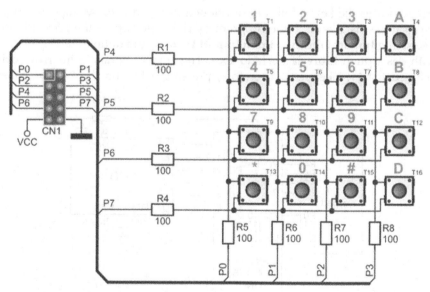

FIG. 10.18   Connection diagram of the keypad.

named as P0, P1, P2, and P3. Notice that 100 ohm resistors are connected in series with every row and column of the keypad.

The keypad is connected to the following PORTC pins of the development board:

Keypad Pin	PORTC Pin
P4	PC_6
P5	PC_7
P6	PC_8
P7	PC_9
P0	PC_10
P1	PC_11
P2	PC_12
P3	PC_13

Fig. 10.19 shows the circuit diagram of the project. The LCD is connected to PORTC of the development board as in the previous project.

10.5.5 The PDL

Fig. 10.20 shows the program PDL.

10.5.6 Program Listing

The program uses the keypad library developed by Yoong Hor Meng (Copyright, 2012). This library can be installed into the program by right clicking on the project folder (**Calc**) and then selecting **Import Library… From Import Wizard**, and double clicking on the keypad library with description **Libreria Teclado 4×4**. The keypad library is interrupt driven and was developed for 4×4 keypads. Therefore, pressing a key on the keypad generates an

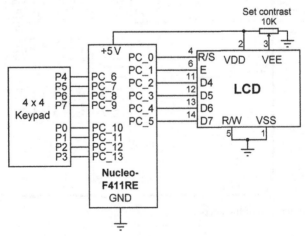

FIG. 10.19　Circuit diagram of the project.

BEGIN/MAIN
    Define LCD to development board interface
    Define keypad to development board interface
    Define key table
    DO FOREVER
        Display heading
        CALL Calculate(1) to read first number
        CALL Calculate(2) to read second number
        Read the required operation
        IF operation is addition **THEN**
            Add two numbers
        ELSE IF operation is subtraction **THEN**
            Subtract two numbers
        ELSE IF operation is multiplication **THEN**
            Multiply two numbers
        ELSE IF operation is division **THEN**
            Divide two numbers
        ENDIF
        Display result
        Wait 5 seconds
    ENDDO
END/MAIN

BEGIN/Calculate (n)
    Display No'n':
    DO FOREVER
        IF character received from keypad **THEN**
            IF character is not Enter **THEN**
                Calculate total number entered
            ELSE
                Exit from loop
            ENDIF
        ENDIF
    ENDDO
    Return total number
END/Calculate

FIG. 10.20 Program PDL.

interrupt where the index of this key (0–15) is available inside the interrupt service routine. A key table must be defined at the beginning of the program with the names of the keys. By using the index we can find which key has been pressed.

The program listing (program: **Calc**) is shown in Fig. 10.21. At the beginning of the program the header files, **mbed.h**, **TextLCD.h**, and **keypad.h** are included in the program. The interface between the LCD and the development board are defined as in the previous project. Also, the interface between the rows and columns of the LCD and the development board are defined. The rows of the keypad are connected to GPIO pins PC_6–PC_9. Similarly, the columns of the keypad are connected to GPIO pins PC_10–PC_13. The keypad characters are stored in an array called **KeyTable**. Inside the main program function **KeyPadISR** is

```
/**************************************************************************

                        KEYPAD BASED CALCULATOR
                        =========================

In this project a HD44780 controller type LCD is connected to PORTC
of the Nucleo-F411RE development board. In addition a 4 x 4 keypad is
conencted to port C. The program is a basic calculator that can do
addition, subtraction, multiplication, and division of integer numbers
only. For example, a typical operation to add numbers 20 and 10 is
as follows (E is the Enter key, i.e. key # on the keypad):

     CALCULATOR    <heading>, for 2 seconds
     No1: 20 E     <column 0, row 0>
     No2: 10 E     <column 0, row 1>

     Op: +   E     <column 0, row 0>
     Res = 30      <column 0, row 1>, for 5 seconds

Author: Dogan Ibrahim
Date  : September 2018
File  : Calc
**************************************************************************/
#include "mbed.h"
#include "TextLCD.h"
#include "keypad.h"

volatile int flag;
unsigned int Index;

TextLCD lcd(PC_0, PC_1, PC_2, PC_3, PC_4, PC_5);        // LCD interface

//
// Keypad interface to the development board
//              r0   r1   r2   r3   c0    c1    c2    c3
Keypad keypad(PC_9,PC_8,PC_7,PC_6,PC_13,PC_12,PC_11,PC_10);

//
// The character table
//
char KeyTable[] =
{
    '1', '2', '3', '+',    // r3
    '4', '5', '6', '-',    // r2
    '7', '8', '9', '*',    // r1
    'C', '0', 'E', '/'     // r0
};
//   c3   c2   c1   c0

//
// This function is called whenever a key is pressed on the keypad. Index
// stores the index number of the key (0 to 15) in KeyTable
```

FIG. 10.21 Program listing.

(Continued)

```
//
unsigned int KeyPadISR(unsigned int index)
{
    Index = index;
    flag = 1;
    return 0;
}


//
// This function clears the LCD and waits for a while until finished
//
void LCDClear()
{
    lcd.cls();
    wait(0.5);
}



//
// This function positions the LCD cursor at the specified row and
// column and waits for a while until finished
//
void LCDCursor(int column, int row)
{
    lcd.locate(column, row);
    wait(0.5);
}


//
// This function displays No1: or No2: and returns the integer number
// entered by te user until the Enter (E) key is pressed on the keypad.
// Variable total is the integer number entered by the user. Variable
// flag is set to 1 when a character is entered on the keypad.
//
int Calculate(int n)
{
    int total = 0;

    lcd.printf("No %d: ", n);                          // No1: or No2:
    while(1)
    {
        if(flag == 1)
        {
            if(KeyTable[Index] != 'E')                 // Is it E?
            {
                total = total*10+KeyTable[Index] - '0';
                lcd.printf("%c",KeyTable[Index]);
                flag = 0;
            }
            else                                       // E is entered
            {
                flag = 0;
                break;                                 // Exit loop
            }
        }
```

FIG. 10.21, CONT'D

(Continued)

```
        }
        return total;                                    // Return total
}


//
// Main program
//
int main()
{
    int no1, no2, res;
    char Op;

    keypad.CallAfterInput(KeyPadISR);                    // Attach int
    keypad.Start();                                      // Start keypad int
    wait(1.0);

  while(1)                                               // Do forever
  {
    LCDClear();                                          // Clear LCD
    lcd.printf("CALCULATOR");                            // Display heading
    wait(2.0);                                           // Wait 2 seconds
    LCDClear();                                          // Clear LCD
    LCDCursor(0, 0);                                     // At 0,0
    no1 = Calculate(1);                                  // Get first no
    LCDCursor(0, 1);                                     // At 0,1
    no2 = Calculate(2);                                  // Get second no

    LCDClear();                                          // Clear LCD
    LCDCursor(0, 0);                                     // At 0,0
    lcd.printf("Op: ");                                  // Display Op:
    while(1)
    {
        if(flag == 1)
        {
            Op = KeyTable[Index];                        // Get Op
            flag=0;
            break;                                       // Exit loop
        }
    }

    switch(Op)                                           // what Op?
    {
        case '+':                                        // If +
            res = no1 + no2;
            break;
        case '-':                                        // If -
            res = no1 - no2;
            break;
        case '*':                                        // If *
            res = no1 * no2;
            break;
        case '/':                                        // If /
            res = no1 / no2;
            break;
    }
```

FIG. 10.21, CONT'D   *(Continued)*

```
        LCDCursor(0, 0);                  // At 0,0
        lcd.printf("%c  ", Op);           // Display Op
        LCDCursor(0, 1);                  // At 0,1
        lcd.printf("Res = %d",res);       // Display res
        wait(5.0);                        // Wait 5 sec
    }
}
```

FIG. 10.21, CONT'D

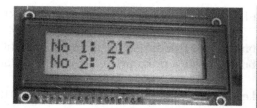 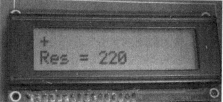

FIG. 10.22 Example displays on the LCD.

attached to keypad interrupts using function **CallAfterInput**, and keypad interrupts are started. The remainder of the program is run in an endless loop. Initially the heading **CAL-CULATOR** is displayed for 2 s. Inside this loop text **No1:** is displayed at column 0, row 1 of the LCD and function **Calculate** is called to read the first integer number from the keypad. Variable **flag** is set to 1 when the user presses a key on the keypad. The number entered must be terminated by the Enter (E) key. Variable **total** stores the integer number entered by the user and this value is returned to the main program. Then text **No2:** is displayed at column 0, row 1 of the LCD and function **Calculate** is called to read the second number from the keypad. Again this number must be terminated by pressing the Enter (**E**) key. The program then displays text **Op:** and waits until the user enters the required operation (+ - * /). The result is then calculated using a **switch** statement and is displayed at column 0, row 1 of the LCD. The program repeats after 5 s of delay. Notice that functions are created for the LCD functions **cls** and **locate** and small delays are used after each function. This was found to be necessary to improve the LCD timing as without these delays it was noticed that the LCD can display unpredictable characters sometimes.

Fig. 10.22 shows example displays from the program.

10.5.7 Suggestions for Additional Work

Modify the project and design a mini electronic organ for one octave using the keypad to play the musical notes. Connect a buzzer to one of the GPIO ports.

Modify the program given in Fig. 10.21 so that it accepts floating point numbers and do floating point calculations.

10.6 SUMMARY

In this chapter we have learned about the following:

- LCD displays
- LCD connection to a Nucleo-F411RE development board
- LCD functions
- Projects using LCDs with the Nucleo-F411RE development board

10.7 EXERCISES

1. Draw the circuit diagram to show how an LCD can be connected to a Nucleo-F411RE development board.
2. An ultrasonic sensor pair is connected to a Nucleo-F411RE development board. In addition an LCD is connected to the development board. Write a program to display the distance from the sensors on the LCD.
3. Draw the circuit diagram to show how a keypad can be connected to a Nucleo-F411RE development board.
4. A keypad and an LCD are connected to a Nucleo-F411RE development board. Write a program to display the numbers entered on the keypad on the LCD.
5. It is required to connect a display to a microcontroller system. Explain why you might have to choose an LCD instead of a 7-segment display.

11

I²C Bus Projects

11.1 OVERVIEW

In this chapter we shall be developing projects using the I²C (interintegrated circuit) bus with the Nucleo-F411RE development board.

The I²C bus was invented by Philips Semiconductor in 1982 for connecting peripheral devices and microcontrollers over short distances. The bus uses two open collector (or open drain) bidirectional lines pulled up with resistors. The SDA is the serial data line and SCL is the serial clock line. Although the bus is bidirectional, data can travel only in one direction at any time. I²C is a bus with 7-bit address space and achieves bus speeds of 100 kbits/s in standard mode and 400 kbits/s in fast mode (faster bus speeds are also available with Version 2.0 of the bus protocol). Devices on the bus can be one or more master nodes and one or more slave nodes. The master nodes initiate the communication and generate the clock signals on the bus. Slave nodes receive the clock signals and respond when addressed by a master.

Fig. 11.1 shows an example I²C system with one master and three slaves. In a typical application the master initiates the communication on the bus by signaling a start condition. This is followed by 7 bits of address information (10-bit addressing mode is also available), and one data direction bit, where a LOW means that the master is writing to the slave, and a HIGH means that the master is reading from the slave. With 7 bits of address up to 128 devices can be connected to the bus. When reading and writing to the bus we have to specify the device address, register address, and the number of bytes.

Mbed supports a number of functions for both master and slave I²C bus communication. In the I²C projects in this book the Nucleo-F411RE development board will be the master and one or more slaves will be connected to the bus. Since we will be programming the master only, a list of the functions available for the master nodes is presented in Table 11.1. When creating an I²C variable we have to specify the GPIO pins for SDA and SCL.

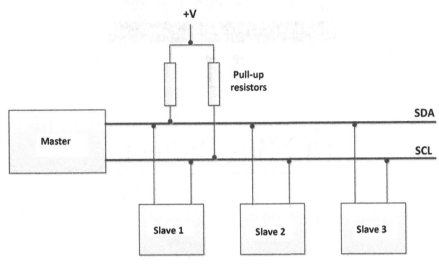

FIG. 11.1   I²C bus with one master and three slaves.

TABLE 11.1   Mbed I²C Master Node Functions

Function	Description
frequency	Set the frequency of communication on the I²C bus
read	Read from a slave
write	Write to a slave
start	Create a start condition on the I²C bus
stop	Create a stop condition on the I²C bus

11.2 NUCLEO-F411RE I²C GPIO PINS

There are three I²C modules on the Nucleo-F411RE development board. The following are the GPIO pins for these modules:

I²C Module	Signal	GPIO Pin
I²C1	SDA	PB_7, PB_9
I²C1	SCL	PB_6, PB_8
I²C2	SDA	PB_3
I²C2	SCL	PB_10
I²C3	SDA	PB_4, PC_9
I²C3	SCL	PA_8

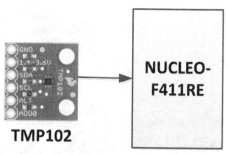

TMP102

FIG. 11.2   Block diagram of the project.

11.3 PROJECT 1—TMP102 TEMPERATURE SENSOR

11.3.1 Description

In this project the TMP102 temperature sensor module is used to read and display the ambient temperature on the PC screen. This sensor is I^2C bus compatible.

11.3.2 Aim

The aim of this project is to show how an I^2C bus compatible device can be connected to the Nucleo-F411RE development board and also how it can be programmed using Mbed.

11.3.3 Block Diagram

The block diagram of the project is shown in Fig. 11.2.

11.3.4 Circuit Diagram

TMP102 module (see Fig. 11.3) has the following pins:

GND: Power supply ground
Vcc: Power supply (1.4–3.6 V)
SDA: I^2C data line
SCL: I^2C clock line
ALT: Alert
ADD0: I^2C address select

The ADD0 pin is used to select the device address. By default the device address is 0x48. Four different addresses can be selected depending on where the ADD0 pin is connected to:

ADD0 Connection	I^2C Address
No connection	0x90 (default)
GND	0x90
Vcc	0x91
SDA	0x92
SCL	0x93

FIG. 11.3   TMP102 module.

Fig. 11.4 shows the circuit diagram of the project. Here, the default device address of 0x90 is selected by leaving the ADD0 pin unconnected. The alert pin is used to generate a signal when the temperature is above or below a limit and this pin is not used in this project. The SDA and SCL pins of the module are connected to I²C1 pins (PB_7, I²C1 SDA, and PB_6, I²C1 SCL) of the development board. Pins Vcc and GND of the module are connected to pins +3.3 V (pin 16, connector CN7) and GND (pin 8, connector CN7) of the development board, respectively. The sensor module has built-in 4.7 K pull up resistors on the SDA and SCL lines as required by the I²C specifications.

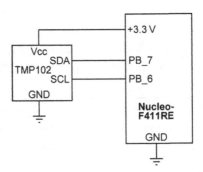

FIG. 11.4   Circuit diagram of the project.

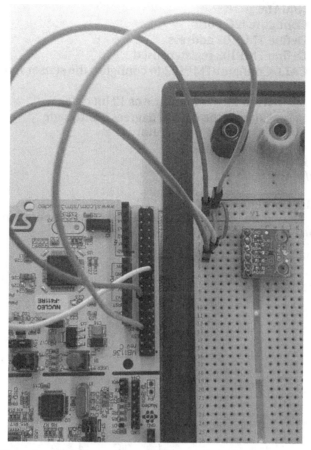

FIG. 11.5 Circuit built on a breadboard.

11.3.5 The Construction

Fig. 11.5 shows the circuit built on a breadboard.

11.3.6 The PDL

Fig. 11.6 shows the program PDL.

11.3.7 Program Listing

TMP102 is a very low-current (10 μA) temperature sensor chip that can be used to measure ambient temperature in the range −40°C to +125°C with a resolution of 0.0625°C and an

BEGIN/MAIN
 Configure I2C ports
 Define TMP102 address
 Define TMP102 registers used
 CALL ConfigureTMP102 to configure the sensor module
 DO FOREVER
  Read the temperature as 12 bit data
  Convert the data to degrees Centigrade
  Display the temperature
  Wait 1 second
 ENDDO
END/MAIN

BEGIN/ConfigureTMP102
 Configure the TMP102 module to normal operation
END/ConfigureTMP102

FIG. 11.6   Program PDL.

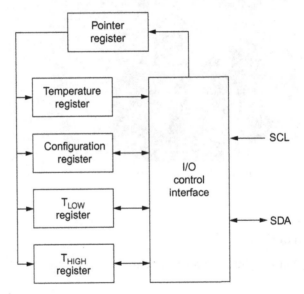

FIG. 11.7   TMP102 registers.

accuracy of ±0.5°C. The device has a 12-bit resolution and is I²C, Two-Wire, and SMBus compatible. TMP102 can have four different device addresses and hence up to four such devices can be connected to the bus. The programming model of the device is shown in Fig. 11.7. The Pointer Register is used to select the register to be configured and it selects the other registers as follows:

00	Select the Temperature Register (read only)
01	Select the Configuration Register
10	Select the T_{LOW} Register
11	Select the T_{HIGH} Register

The Configuration Register is a 16-bit register and after power up or reset it is necessary to configure this register. The register is configured by sending two consecutive bytes to it. The following options can be configured (see the TMP102 data sheet for details: http://www.ti.com/lit/ds/symlink/tmp102.pdf):

- shutdown mode
- thermostat mode
- polarity
- fault queue
- conversion resolution
- one-shot mode
- extended conversion mode
- alert bit
- conversion rate

In this example we will configure this register with the normal operating values by sending 0x60 followed by 0xA0. These are also the default register values after power up or reset. 0x60 sets the conversion resolution to 12 bits, continuous conversion mode. 0xA0 sets nonextended mode of operation and 4 Hz conversion rate.

The Temperature Register returns the temperature in two bytes as shown in Fig. 11.8, where 4 bits of the lower byte are cleared to 0.

The temperature can be extracted from these two bytes by the following steps:

For positive temperatures:

- Shift the HIGH byte by 4 bits to the left into a 16-bit word called say, M.
- Shift the LOW byte by 4 bits to the right into a byte called say, L.
- Add M and L to find the 16-bit result.
- Multiply the result with 0.0625 (resolution) to find the temperature.

D7	D6	D5	D4	D3	D2	D1	D0
T11	T10	T9	T8	T7	T6	T5	T4
(T12)	(T11)	(T10)	(T9)	(T8)	(T7)	(T6)	(T5)

D7	D6	D5	D4	D3	D2	D1	D0
T3	T2	T1	T0	0	0	0	0
(T4)	(T3)	(T2)	(T1)	(T0)	(0)	(0)	(1)

FIG. 11.8 The Temperature Register.

For negative temperatures:

- Shift the HIGH byte by 4 bits to the left into a 16-bit word called say, M.
- Shift the LOW byte by 4 bits to the right into a byte called say, L.
- Add M and L to find the 16-bit result.
- Generate 2's complement of the result (complement and add 1).
- Multiply the result with 0.0625 (resolution) to find the temperature.

For example, if the returned two bytes are: 0011 0010 1000 0000

M = 0011 0010 0000
L = 1000
M + L = 0011 0010 1000 which is equivalent to 808 in decimal. Therefore, 801 × 0.0625 = 50.5°C.

Similarly, for a negative temperature, if the returned bytes are: 1110 0111 1000 0000

M = 1110 0111 0000
L = 1000
M + L = 1110 0111 1000

Taking the complement and adding 1 (2's complement) we get: 0001 1000 0111 + 1 = 0001 1000 1000 which is equivalent to 392. Therefore 392 × 0.0625 = 24.5°C, or −24.5°C.

Fig. 11.9 shows the program listing (program: **TMP102**). At the beginning of the program GPIO pins PB_7 and PB_6 are assigned to I²C1 with the name **TMP102**, the device address is set to 0x90, and the registers used in the program are defined. Inside the main program function **ConfigureTMP102** is called to configure the sensor to normal mode of operation. The remainder of the program runs in an endless loop. Inside this loop the temperature is read as a 12-bit data and is converted into a 16-bit data in variable **M**. The actual temperature in °C is then calculated and displayed on the PC screen in the format: **nn.nn**. Notice that only positive temperatures are displayed by this program. The loop is repeated every second.

A typical display of the temperature is shown in Fig. 11.10.

11.3.8 Suggestions for Additional Work

Modify the program given in Fig. 11.9 so that the negative temperatures can also be displayed.

11.4 SUMMARY

In this chapter we have learned about the following:

- I²C bus
- Mbed I²C bus functions
- A project using an I²C compatible device

```
/**********************************************************************

                    TMP102 TEMPERATURE SENSOR
                    =========================

This is an I2C based temperature project. In this project a TMP102 type
temperature sensor module is connected to I2C1 port of the Nucleo-F411RE
development board. The program reads and displays the ambient temperature
on the PC screen.

Author: Dogan Ibrahim
Date  : August 2018
File  : TMP102
***********************************************************************/
#include "mbed.h"
I2C TMP102(PB_7, PB_6);                         // I2C SDA, SCL
Serial MyPC(USBTX, USBRX);                       // Serial TX, RX

const int TMP102Address = 0x90;                  // TMP102 address
char ConfigRegister[3];                          // Config register
char TemperatureRegister[2];                     // Temperature Register
float Temperature;

//
// This function configures the TMP102 after power up or reset
//
void ConfigureTMP102()
{
    ConfigRegister[0] = 0x01;                    // Point to Config Register
    ConfigRegister[1] = 0x60;                    // Upper byte
    ConfigRegister[2] = 0xA0;                    // Lower byte
    TMP102.write(TMP102Address, ConfigRegister, 3); // Write 3 bytes
}


int main()
{
    unsigned short M;
    char L;

    ConfigureTMP102();                           // Configure TMP102
    ConfigRegister[0] = 0x00;                    // Point to Temp Register
    TMP102.write(TMP102Address, ConfigRegister, 1); // Write 1 byte

//
// Read and display the ambient temperature on the PC screen every second
//
    while(1)
    {
        TMP102.read(TMP102Address, TemperatureRegister, 2);
        M = TemperatureRegister[0] << 4;
        L = TemperatureRegister[1] >> 4;
        M = M + L;
        Temperature = 0.0625 * M;
        MyPC.printf("\n\rTemperature = %5.2f", Temperature);
        wait(1.0);
    }
}
```

FIG. 11.9 Program listing.

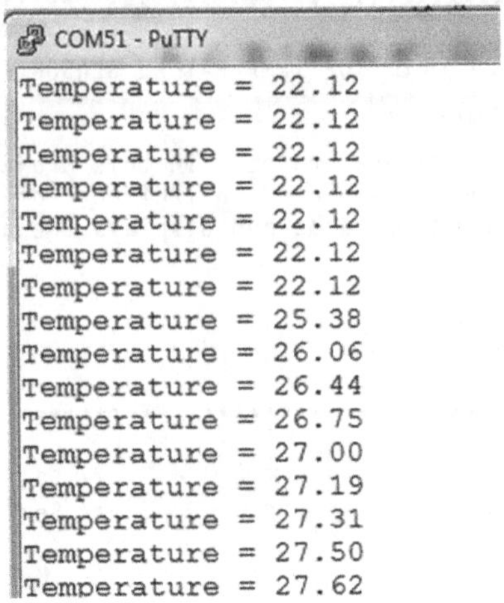

```
COM51 - PuTTY
Temperature = 22.12
Temperature = 22.12
Temperature = 22.12
Temperature = 22.12
Temperature = 22.12
Temperature = 22.12
Temperature = 22.12
Temperature = 25.38
Temperature = 26.06
Temperature = 26.44
Temperature = 26.75
Temperature = 27.00
Temperature = 27.19
Temperature = 27.31
Temperature = 27.50
Temperature = 27.62
```

FIG. 11.10   Typical display of the temperature.

11.5 EXERCISES

1. Explain how the I²C bus operates.
2. Which pins can be used for the I²C interface on a Nucleo-F411RE development board.
3. A TMP102 temperature sensor chip is connected to the Nucleo-F411RE development board. Additionally an LCD is connected to the development board. Write a program to display the ambient temperature every minute.

CHAPTER

12

SPI Bus Projects

12.1 OVERVIEW

In this chapter we shall be developing projects using the SPI (Serial Peripheral Interface) bus with the Nucleo-F411RE development board.

The SPI bus is one of the commonly used protocols to connect sensors and many other devices to microcontrollers. The SPI bus is a master-slave-type bus protocol. In this protocol, one device (the microcontroller) is designated as the master, and one or more other devices (usually sensors) are designated as slaves. In a minimum bus configuration there is one master and only one slave. The master establishes communication with the slaves and controls all the activity on the bus.

Fig. 12.1 shows an SPI bus example with one master and three slaves. The SPI bus uses three signals: clock (SCK), data in (SDI), and data out (SDO). The SDO of the master is connected to the SDIs of the slaves, and SDOs of the slaves are connected to the SDI of the master. The master generates the SCK signals to enable data to be transferred on the bus. In every clock pulse 1 bit of data is moved from master to slave, or from slave to master. The communication is only between a master and a slave, and the slaves cannot communicate with each other. It is important to note that only one slave can be active at any time since there

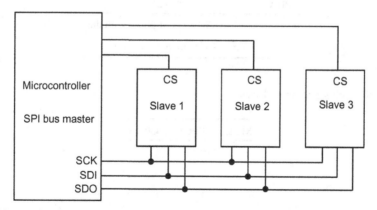

FIG. 12.1   SPI bus with one master and three slaves.

ARM-based Microcontroller Projects Using mbed
https://doi.org/10.1016/B978-0-08-102969-5.00012-4

TABLE 12.1 Mbed SPI Master Node Functions

Function	Description
format	Configure the data transmission format
frequency	Set the SPI bus frequency in Hz
write	Write to a slave on the SPI bus and return response

is no mechanism to identify the slaves. Thus, slave devices have enable lines (e.g., CS) which are normally controlled by the master. A typical communication between a master and several slaves is as follows:

- Master enables slave 1.
- Master sends SCK signals to read or write data to slave 1.
- Master disables slave 1 and enables slave 2.
- Master sends SCK signals to read or write data to slave 2.
- The above process continues as required.

The SPI signal names are also called MISO (Master in, Slave out), and MOSI (Master out, Slave in). Clock signal SCK is also called SCLK and the CS is also called SSEL. Mbed supports a number of functions for both master and slave SPI bus communication. In the SPI projects in this book the Nucleo-F411RE development board will be the master and one or more slaves will be connected to the bus. Since we will be programming the master only, a list of the functions available for the master nodes is presented in Table 12.1. The **format** function takes two arguments: data length and mode. The data length is usually 8 bits. The mode can have four values depending on the required clock polarity (CPOL) and clock phase (CPHA). CPOL and CPHA can have the following values:

CPOL	Clock Active State
0	Clock active HIGH
1	Clock active LOW
CPHA	**Clock Phase**
0	Clock out of phase with data
1	Clock in phase with data

The four SPI modes are as follows:

Mode	CPOL	CPHA
0	0	0
1	0	1
2	1	0
3	1	1

When creating an SPI bus variable we have to specify the GPIO pins for the MOSI, MISO, and SCLK. The default bus speed is 1 MHz (1,000,000 Hz), default data length is 8 bits, and the default mode is 0. The mode depends on the requirements of the slave device and the slave data sheet should be checked before a mode is selected.

12.2 NUCLEO-F411RE SPI GPIO PINS

There are five SPI modules on the Nucleo-F411RE development board. The following are the GPIO pins for these modules:

SPI Module	Signal	GPIO Pin
SPI1	SSEL	PA_15
SPI1	SCLK	PA_5
SPI1	MISO	PA_6
SPI1	MOSI	PA_7
SPI2	MISO	PC_2, PB_14
SPI2	MOSI	PC_3, PB_15
SPI2	SSEL	PB_9, PB_12
SPI2	SCLK	PC_7, PB_13
SPI3	MOSI	PC_12, PB_5
SPI3	SCLK	PC_10, PB_3
SPI3	MISO	PC_11, PB_4
SPI3	SSEL	PA_4
SPI4	MOSI	PA_1
SPI4	MISO	PA_11
SPI5	SCLK	PB_0
SPI5	MISO	PA_12
SPI5	MOSI	PB_8, PA_10
SPI5	SSEL	PB_1

12.3 PROJECT 1—GENERATING SQUARE WAVE

12.3.1 Description

In this project a DAC (digital-to-analog converter) chip is used to generate a square wave signal with a frequency of 1 kHz (period = 1 ms), 50% duty cycle, and an amplitude of 1 V. The DAC used is SPI bus compatible.

12.3.2 Aim

The aim of this project is to show how an SPI bus compatible device can be connected to the Nucleo-F411RE development board and also how it can be programmed using Mbed.

12.3.3 Block Diagram

The block diagram of the project is shown in Fig. 12.2.

12.3.4 Circuit Diagram

In this project the MCP4921-type SPI bus compatible DAC chip is used. MCP4921 is a 12-bit serial DAC manufactured by Microchip Inc., having the following basic specifications:

- 12-bit resolution
- up to 20 MHz clock rate (SPI)
- fast settling time of 4.5 μs
- unity or 2× gain output
- external V_{ref} input
- 2.7–5.5 V operation
- extended temperature range (−40°C to +125°C)
- 8-pin DIL package

Fig. 12.3 shows the pin layout of the MCP4921. The pin definitions are as follows:

V_{DD}, AV_{SS}: power supply and ground
CS: chip select (LOW to enable the chip)
SCK, SDI: SPI bus clock and data in
V_{OUTA}: analog output
V_{REFA}: reference input voltage
LDAC: DAC input latch (transfers the input data to the DAC registers.

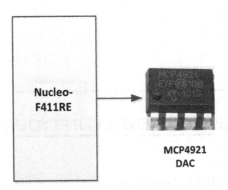

FIG. 12.2   Block diagram of the project.

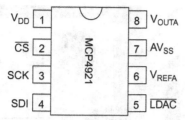

FIG. 12.3 Pin layout of MCP4921 DAC.

Normally tied to ground so that CS controls the data transfer.

Fig. 12.4 shows the circuit diagram of the project. SPI2 module of the development board is used to interface with the DAC chip, where MOSI is pin PC_3 and SCLK is pin PC_7. The CS input of the DAC chip is connected to PC_0 of the development board. The output of the DAC is connected to a PC-based oscilloscope in order to record the generated waveform.

12.3.5 The Construction

The project was constructed on a breadboard as shown in Fig. 12.5 and connections were made to the development board using jumper wires.

12.3.6 The PDL

Fig. 12.6 shows the program PDL.

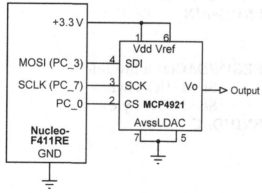

FIG. 12.4 Circuit diagram of the project.

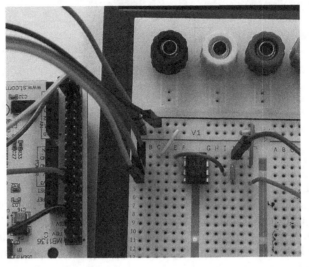

FIG. 12.5   Project constructed on a breadboard.

BEGIN/MAIN
      Define the SPI bus interface
      Set signal amplitude to 1V
      DO FOREVER
            CALL DAC (amplitude)
            Wait 0.5ms
            CALL DAC (0V)
            Wait 0.5ms
      ENDDO
END/MAIN

BEGIN/DAC(amplitude)
      Send HIGH byte
      Send LOW byte
END/DAC

FIG. 12.6   Program PDL.

12.3.7 The Program Listing

The digital input of a DAC converter can either be in serial or parallel form. In a parallel converters the width of digital input is equal to the width of the converter. For example, a 12-bit converter has 12 input bits. Serial converters in general use the SPI or the I^2C bus and basically a clock and a data signal are used to send data to the converter. Parallel converters provide much faster conversion times but they are housed in larger packages. The DACs are manufactured as either unipolar or bipolar as far as the output voltages are concerned. Unipolar converters can provide only positive output voltages, whereas bipolar converters provide both positive and negative voltages. In this book we will be using only unipolar converters.

The relationship between the digital input-output and the voltage reference is given by

$V_o = DV_{ref}/2^n$

where V_o is the output voltage, V_{ref} the reference voltage, D the digital data, and n is the width of the converter. For example, in a 12-bit converter (resolution $=$ 12-bits) with $+3.3V$ reference voltage,

$V_o = 3.3D/2^{12} = 0.805\,D\,mV$

Thus, for example, if the input digital value is 1, the analog output voltage will be $0.805\,mV$, if the input value is 2, the analog output voltage will be $1.61\,mV$, and so on.

The program listing (program: **SPIDAC**) of the project is shown in Fig. 12.7. At the beginning of the program SPI channel 2 is assigned to variable **dac** and PC_0 (**CS** of the DAC chip) is configured as an output. Inside the main program, assuming a reference voltage of $3.3V$, variable amplitude is set to $1V$ ($1000\,mV$) and the DAC chip is disabled. The remainder of the program runs in an endless loop. Inside this loop function **DAC** is called and variable **amplitude** is sent as an argument to the function so that $1V$ analog signal is output from the DAC. Then $0.5\,ms$ delay is inserted and 0 is sent to the DAC chip with again $0.5\,ms$ delay so that the Duty Cycle of the signal is 50%. The loop is repeated until stopped by the user.

Function **DAC** writes data to the DAC chip. Data is written to the DAC in two bytes. The HIGH byte is sent first followed by the LOW byte. The lower byte specifies bits D0:D8 of the digital input data. The upper byte consists of the following bits:

D8:D11: bits D8:D11 of the digital input data
SHDN: 1, output power down mode; 0, disable output buffer
GA: output gain control. 0, gain is $2\times$; 1, gain is $1\times$
BUF: 0, input unbuffered; 1, input buffered
A/B: 0, write to DAC_A; 1, write to DAC_B (MCP4921 supports only DAC_A)

Data is written to the DAC chip over the SPI bus using function **dac.write**.
Fig. 12.8 shows the generated output waveform on the oscilloscope.

12.3.8 Suggestions for Additional Work

Modify the program shown in Fig. 12.6 to generate a sawtooth waveform.

```
/***************************************************************************

                        SPI BUS SQUARE WAVE GENERATOR
                        ==============================

         This is an SPI bus project. An MCP4921 DAC is connected to SPI channel
         2 of the Nucleo-F411RE development board. The program generates analog
         square wave signal through the DAC with frequency of 1kHz (period=1ms)
         and amplitude of 1V.

         Author: Dogan Ibrahim
         Date  : September 2018
         File  : SPIDAC
***************************************************************************/
#include "mbed.h"

SPI dac( PC_3, PC_2, PC_7, PB_9);                   // MOSI,MISO,SCLK,SSEL
DigitalOut CS(PC_0);                                // CS of the DAC chip

//
// This function sends 12-bit digital data to the MCP4921 DAC. The
// HIGH byte is sent first, followed by the LOW byte
//
void DAC(unsigned int value)
{
    char temp;

    CS = 0;                                         // Enable CS
//
// Send HIGH byte
//
    temp = (value >> 8) & 0x0F;                     // Bits 8-11 in temp
    temp |= 0x30;                                   // Define DAC gain=1x
    dac.write(temp);                                // Send HIGH byte
//
// Send LOW byte
//
    temp = value;
    dac.write(temp);                                // Send LOW byte
    CS = 1;                                         // Disable CS
}


//
// Main Program
// Generate square wave with amplitude 1V, and period 1ms
//
int main()
{
    unsigned int amplitude;

    amplitude = 1000 * 4095 / 3300;                 // 1V
    CS = 1;                                         // Disable CS

    while(1)
    {
        DAC(amplitude);                             // Send 1V to DAC
        wait(0.0005);                               // Wait 0.5ms
        DAC(0);                                     // Send 0V to DAC
        wait(0.0005);                               // Wait 0.5ms
    }
}
```

FIG. 12.7 Program listing.

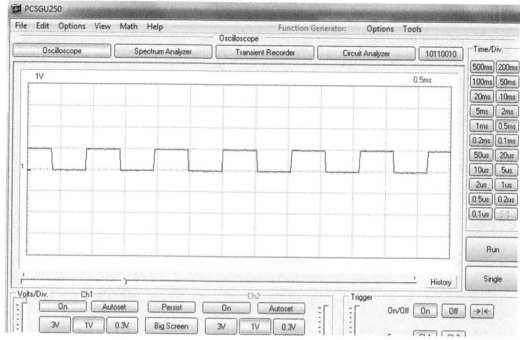

FIG. 12.8   Output waveform.

12.4 SUMMARY

In this chapter we have learned about the following:

- SPI bus
- Mbed SPI bus functions
- A project using an SPI compatible device

12.5 EXERCISES

1. Explain the signals used in the SPI bus.
2. Draw a table comparing the advantages and disadvantages of using SPI bus instead of I²C bus.
3. The Nucleo-F411RE development board is to be connected to three SPI bus slave sensors. Draw the circuit diagram of this project.
4. What are the limitations for connecting devices to the SPI bus?
5. Explain the Mbed functions that can be used in SPI bus-based projects.

CHAPTER

13

UART Projects

13.1 OVERVIEW

In this chapter we shall be developing projects using the serial communication module UART with the Nucleo-F411RE development board.

Serial communication is a simple means of sending data to long distances quickly and reliably. The most commonly used serial communication method is based on the RS232 standard. In this standard data is sent over a single line from a transmitting device to a receiving device in bit serial format at a prespecified speed, also known as the Baud rate, or the number of bits sent each second. Typical Baud rates are 4800, 9600, 19200, 38400, etc.

RS232 serial communication is a form of asynchronous data transmission where data is sent character by character. Each character is preceded with a Start bit, seven or eight data bits, an optional parity bit (check bit), and one or more stop bits. The most commonly used format is eight data bits, no parity bit, and one stop bit. The least significant data bit is transmitted first, followed by the other bits, and the most significant bit transmitted last.

Data in serial asynchronous communication can either be sent using the standard RS232 protocol or two devices can simply be connected together provided they satisfy the logic gate interface rules (e.g., TTL to TTL, or TTL to CMOS, etc.). In the RS232 protocol, a logic HIGH is defined to be at −12 V, and a logic 0 is at +12 V. Fig. 13.1 shows how character **A** (ASCII binary pattern 0010 0001) is transmitted over a serial line via the RS232 protocol. The line is normally idle at −12 V. The start bit is first sent by the line going from HIGH to LOW. Then eight data bits are sent starting from the least significant bit. Finally the stop bit is sent by raising the line from LOW to HIGH.

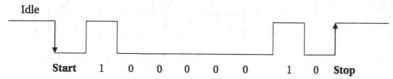

FIG. 13.1 Sending character "A" in serial format.

In serial connection a minimum of three lines are used for communication: transmit (TX), receive (RX), and ground (GND). Two devices are used in serial communication: the transmitter, and the receiver. The devices are connected such that the TX of one device is connected to the RX of the other device, and its RX is connected to the TX. Some high-speed serial communication systems use additional control signals for synchronization, such as CTS, DTR, RTS, and so on. Some systems use software synchronization techniques where a special character (XOFF) is used to tell the sender to stop sending, and another character (XON) is used to tell the sender to restart transmission. In this chapter we will be using low-speed communication and therefore the basic pins presented in Table 13.1 will be used with no hardware or software synchronization.

Serial devices are connected to each other using two types of connectors: 9-way connector, or 25-way connector. Table 13.1 illustrates the TX, RX, and GND pins of each types of connectors. The connectors used in RS232 serial communication are shown in Fig. 13.2.

As described above, RS232 voltage levels are at ±12 V. On the other hand, microcontroller input-output ports operate at 0 to +3.3 V or 0 to +5 V voltage levels. It is therefore necessary to translate the voltage levels before a microcontroller can be connected to an RS232 compatible device. Thus, the output signal from the microcontroller has to be converted into ±12 V, and the input from an RS232 device must be converted into 0 to +5 V before it can be connected to a microcontroller. This voltage translation is normally done using special RS232 voltage converter chips. One such popular chip is the MAX232. In the UART project in this chapter

TABLE 13.1   Minimum Required Pins for Serial Communication

Pin	Function
9-PIN CONNECTOR	
2	Transmit (TX)
3	Receive (RX)
5	Ground (GND)
25-PIN CONNECTOR	
2	Transmit (TX)
3	Receive (RX)
7	Ground (GND)

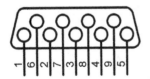

9-Way connecter

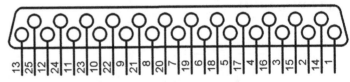

25-Way connecter

FIG. 13.2   RS232 connectors.

the RS232 protocol is not used and the two communicating devices are connected to each other directly.

Table 13.2 illustrates the functions supported by Mbed UART.

TABLE 13.2   Mbed UART Functions

Function	Description
baud	Set the serial communication baud rate
format	Set the communication format
getc	Read a character from serial port
putc	Write a character to serial port
printf	Write formatted string to serial port
scanf	Read formatted string from serial port
readable	Determine if a there a character available to read
writeable	Determine if there is space to write a character
attach	Attach an interrupt service routine function to serial port

13.2 NUCLEO-F411RE UART GPIO PINS

There are three UART modules on the Nucleo-F411RE development board. The following are the GPIO pins for these modules:

UART Module	Signal	GPIO Pin
UART1	TX	PA_15, PB_6, PA_9
UART1	RX	PB_7, PB_3, PA_10
UART1	CTS	PA_11
UART2	TX	PA_2 (**PC communication**)
UART2	RX	PA_3 (**PC communication**)
UART2	CTS	PA_0
UART2	RTS	PA_1
UART6	RX	PC_7
UART6	TX	PC_6
UART6	RTS	PA_12

Notice that UART2 is used to communicate with the PC and is not freely available as a general purpose serial communication module.

13.3 PROJECT 1—TWO NUCLEO BOARDS COMMUNICATING THROUGH UART

13.3.1 Description

In this project two Nucleo-F411RE development boards called **Sender Node** and **Receiver Node** are connected through UART interface. A temperature sensor is connected to the **Sender Node**. This board reads the ambient temperature and sends it to the **Receiver Node** every second where it is displayed on the PC screen.

13.3.2 Aim

The aim of this project is to show how the UART module can be used on the Nucleo-F411RE development board and also how it can be programmed using Mbed.

13.3.3 Block Diagram

The block diagram of the project is shown in Fig. 13.3 where two Nucleo-F411RE development boards are connected to each other. The temperature sensor is connected to the **Sender Node** and temperature readings are displayed on the PC screen connected to the **Receiver Node**.

13.3.4 Circuit Diagram

Fig. 13.4 shows the circuit diagram of the project. SDA and SCL pins of the TMP102 temperature sensor is connected to pins PB_7 and PB_6 of the **Sender Node**, respectively. UART6 TX pin (PC_6) of the **Sender Node** is connected to UART6 RX pin (PC_7) of the **Receiver Node**.

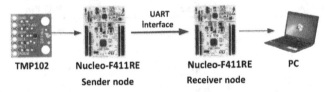

FIG. 13.3   Block diagram of the project.

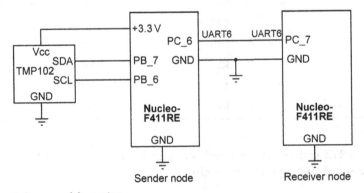

FIG. 13.4   Circuit diagram of the project.

SERVER NODE

BEGIN

      Configure UART6 and assign TX, RX to variable Sender
      Configure TMP102 module
      Set UART6 baud rate to 9600
      DO FOREVER
            Read the ambient temperature
            Convert into Degrees Centigrade
            Wait 1 second
      ENDDO
END

RECEIVER NODE

BEGIN

      Configure UART6 and assign TX, RX to variable Receiver
      Set UART6 baud rate to 9600
      DO FOREVER
            IF UART data is available **THEN**
                  Read the data
                  Display the data on PC screen
            ENDIF
      ENDDO
END

FIG. 13.5 Program PDL.

13.3.5 The PDL

Fig. 13.5 shows the program PDL.

13.3.6 Program Listing

Fig. 13.6 shows the **Sender Node** program listing (program: **Sender**). This program uses the I^2C compatible TMP102 temperature sensor module as was discussed in Chapter 11. At the beginning of the program variable **Sender** is assigned to UART6 pins PC_6 (TX pin) and PC_7 (RX pin). The program reads the ambient temperature every second and sends it to the **Receiver Node** over the UART interface using function **Sender.printf**. In this project, UART6 of the Nucleo board is used to send the temperature data. The baud rate of the communication is set to 9600.

The **Receiver Node** program listing (program: **Receiver**) is shown in Fig. 13.7. At the beginning of the program variable **Receiver** is assigned to UART6 pins PC_6 and PC_7.

This program reads the temperature data using UART6 again at 9600 baud. Function **receiver. scanf** is used to read the data from the UART. The received data is displayed on the PC monitor connected to the **Receiver Node** using function **pc.printf**.

A typical display from the Receiver Node is shown in Fig. 13.8.

```
/*******************************************************************

                        UART - SENDER NODE
                        ===================

This is an I2C based temperature project. In this project a TMP102 type
temperature sensor is connected to a Nucleo-F411RE development board (
called the Sender Node). The program reads teh ambient temperature and
then sends it to the Receiver Node through the UART6 module of the
development board. The baud rate is set to 9600.

Author: Dogan Ibrahim
Date  : September 2018
File  : Sender
*******************************************************************/
#include "mbed.h"
I2C TMP102(PB_7, PB_6);                      // I2C SDA, SCL
Serial Sender(PC_6, PC_7);                   // UART6 TX, RX

const int TMP102Address = 0x90;              // TMP102 address
char ConfigRegister[3];                      // Config register
char TemperatureRegister[2];                 // Temperature Register
float Temperature;

//
// This function configures the TMP102 after power up or reset
//
void ConfigureTMP102()
{
    ConfigRegister[0] = 0x01;                // Point to Config Register
    ConfigRegister[1] = 0x60;                // Upper byte
    ConfigRegister[2] = 0xA0;                // Lower byte
    TMP102.write(TMP102Address, ConfigRegister, 3); // Write 3 bytes
}


int main()
{
    unsigned short M;
    char L;

    ConfigureTMP102();                            // Configure TMP102
    ConfigRegister[0] = 0x00;                     // Point to Temp Register
    TMP102.write(TMP102Address, ConfigRegister, 1); // Write 1 byte
    Sender.baud(9600);                            // Sender baud rate
```

FIG. 13.6 Sender Node program listing.

(Continued)

```
//
// Read the ambient temperature, convert into degrees Centigrade and then
// send it to the Receiver Node over the UART
//
    while(1)
    {
        TMP102.read(TMP102Address, TemperatureRegister, 2);
        M = TemperatureRegister[0] << 4;
        L = TemperatureRegister[1] >> 4;
        M = M + L;
        Temperature = 0.0625 * M;
        Sender.printf("%f\n\r", Temperature);
        wait(5.0);
    }
}
```

FIG. 13.6, CONT'D

```
/**************************************************************************

                        UART - RECEIVER NODE
                        ====================

In this project the Nucleo-F411RE development board (called the Receiver
Node)receives the ambient temperature readings from teh Sender Node and
then displays them on the PC screen. UART6 module of the development
board is used withe the baud rate set to 9600.

Author: Dogan Ibrahim
Date  : September 2018
File  : Receiver
**************************************************************************/
#include "mbed.h"

Serial pc(USBTX, USBRX);                            // PC interface
Serial Receiver(PC_6, PC_7);                        // UART6 TX, RX


int main()
{
    float Temperature;
    Receiver.baud(9600);                            // Sender baud rate
//
// Read the ambient temperature from the Sender Node and then display
// on the PC screen. First check to make sure that data is available
// before attempting to read from UART
//
    while(1)
    {
        if(Receiver.readable())
        {
            Receiver.scanf("%f", &Temperature);
            pc.printf("\n\rTemperature = %5.2f", Temperature);
        }
    }
}
```

FIG. 13.7 Receiver Node program listing.

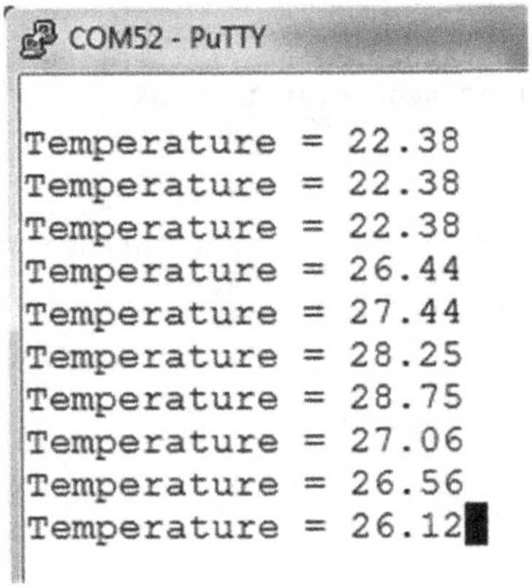

```
COM52 - PuTTY
Temperature = 22.38
Temperature = 22.38
Temperature = 22.38
Temperature = 26.44
Temperature = 27.44
Temperature = 28.25
Temperature = 28.75
Temperature = 27.06
Temperature = 26.56
Temperature = 26.12
```

FIG. 13.8   Typical display from the Receiver Node.

13.4 SUMMARY

In this chapter we have learned about the following:

- UART
- Nucleo-F411RE development board UART modules
- Mbed UART functions
- A project where two Nucleo-F411RE development boards are connected together through one of their UART ports and they communicate with each other.

13.5 EXERCISES

1. What does the acronym UART stand for?
2. Explain the minimum signals that can be used when two devices want to communicate using UARTs.
3. What is the baud rate? What are the typical baud rates?
4. Two Nucleo-F411RE development boards are to be connected via the UART interface. Write a program to show how the data sent by one device can be received and displayed on the PC screen by the other device.
5. What are USBTX and USBRX?
6. Explain how sensor data can be received by a Nucleo-F411RE development board and then displayed on the PC screen. Give an example project with the code.

14

Advanced Projects

14.1 OVERVIEW

In this chapter we shall be developing more complex projects using some of the commonly used communication products such as Wi-Fi and Bluetooth. We shall be using an Android-based smartphone to communicate with our Nuclo-F411RE development board.

14.2 WI-FI

Wi-Fi provides wireless connectivity to your computers, smartphones, PDAs, and ipads. Wi-Fi is also known as the 802.11 network standard and its major advantage is that it is compatible with almost every operating system, printer, or gaming devices. Wi-Fi connectivity is established by using a network router. Nearly all homes in developed countries have network routers which provide users the capabilities to access the internet or their local smart devices such as wireless printers, smartphones, etc.

The Wi-Fi technology has developed through several standards. 802.11a operates at 5 GHz and it offers a maximum of 54 Mbits of data per second. 802.11b operates at 2.4 GHz and offers a much lower speed of 11 Mbits/s. 802.11g is a newer and faster standard based on 2.4 GHz with a maximum speed of 54 Mbits/s. 802.11n operates with separate transmitter and receiver antennas and offers up to 140 Mbits of data per second while operating at the frequency of 5 GHz. Some newer standards such as the 802.11ac uses three antennas and offers data speeds in excess of GBits per second while operating at 5 GHz.

14.3 PROJECT 1—ANDROID—NUCLEO BOARD COMMUNICATION USING THE WI-FI EXPANSION BOARD

14.3.1 Description

In this project the Nucleo compatible Wi-Fi expansion board is plugged on top of the Nucleo-F411RE development board. Additionally, four relays are connected to the development board. The relays are controlled (turned ON or OFF) from an Android mobile phone (or any other device with network access, e.g., PC or tablet) by sending TCP-based commands to the Wi-Fi expansion board through a Wi-Fi link established using a network router.

14.3.2 Aim

The aim of this project is to show how a number of relays connected to the Nucleo-F411RE development board can be controlled wirelessly through a Wi-Fi link using the Nucleo Wi-Fi expansion board. Additionally, the project shows how a TCP-based program can be developed on the Nucleo development board in order to communicate with a TCP application running on an Android mobile phone.

14.3.3 Block Diagram

There are many different Nucleo expansion boards that can be used to simplify the design of projects using the Nucleo development boards. Some expansion boards include sensors such as accelerometer, temperature and humidity sensor, magnetometer, NFC card reader, etc., some include indicators such as LEDs, some include actuators such as DC motor drivers or stepper motor drivers, and some others include Wi-Fi or Bluetooth communication modules. The expansion boards plug on top of the Nucleo development boards through the Arduino connectors.

The block diagram of the project is shown in Fig. 14.1. Here, a Wi-Fi router provides the wireless connection between the mobile phone and the Wi-Fi expansion board. A 4-channel relay board is connected to the Nucleo-F411RE development board. Various indicators or actuators can be connected to these relays.

14.3.4 TCP and UDP Communications

When devices are connected to each other wirelessly through a Wi-Fi router, communication between these devices can be established using either the TCP or the UDP protocols.

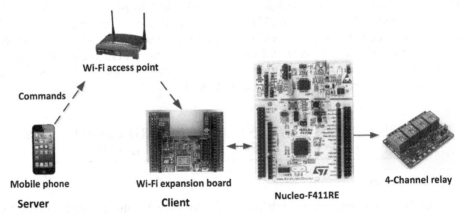

FIG. 14.1   Block diagram of the project.

Both of these protocols are server-client (or master-slave) based where normally one node initiates the transfer request and then both nodes can send or receive data from each other. Data is sent in the form of packets where protocol data are added to the actual data to form a data packet.

The TCP is a reliable connection-based protocol where the communicating nodes must connect to each other before any data transfer can take place. The UDP on the other hand does not require the communicating nodes to connect to each other. The TCP has a bigger packet overhead and as a result it is slower than UDP. Data packets sent by TCP are checked for errors by the receiving nodes and any packets with errors are retransmitted. There is no retransmission in UDP even if a packet contains errors. The TCP packets are acknowledged by the receiving nodes and this guarantees the safe and reliable delivery of the packets to their destinations. The UDP packets on the other hand are not acknowledged and it is possible for packets to be lost. As a summary, TCP protocol should be used if reliable transfer of data packets is important. The UDP protocol on the other hand should be used if it is required to deliver the data as quickly as possible and where the reliability of the data transfer is not so important (e.g., in repetitive data transfer applications where the data content does not change). Examples of programs using the TCP protocol are HTTP, HTTPS, FTP, etc. Example programs using the UDP protocol are DHCP, TFTP, etc.

In this project the Wi-Fi expansion board is the **Client** and the mobile device is the **Server**.

14.3.5 The Nucleo Wi-Fi Expansion Board

The Nucleo Wi-Fi expansion board (X-NUCLEO-IDW01M1) is based on the **SPWF01SA** chip and it provides Wi-Fi capability to the Nucleo development boards (see Fig. 14.2). This board supports up to eight TCP/UDP sockets, dynamic web pages, socket server, and socket client with TLS/SSL encryption. In addition to the Wi-Fi functions, the expansion board also supports 16 GPIOs. The board is controlled with the standard AT commands and is interfaced to the host processor through an UART module.

The Wi-Fi expansion board includes a number of configuration jumpers and LEDs. Two buttons and four LEDs are provided on the board. One of the LEDs indicate the presence

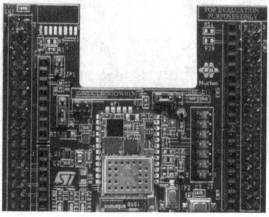

FIG. 14.2 Nucleo Wi-Fi expansion board.

of the power supply, while the three other LEDs are connected to three GPIO ports. Further information about the Wi-Fi expansion board can be obtained from the following website:

https://www.st.com/en/evaluation-tools/stm32-nucleo-expansion-boards.html?querycriteria=productId=SC1971

14.3.6 The 4-Channel Relay Board

In this project a 4-channel relay board (see Fig. 14.3) from Elegoo (www.elegoo.com) is used. This is an opto-coupled relay board having four inputs, one for each channel. The relay inputs are at the bottom right-hand side of the board while the relay outputs are located at the top side of the board. The middle position of each relay is the common point, the connection to its left is the normally closed (NC) contact, while the connection to the right is the normally open (NO) contact. The relay contacts support AC250V at 10A and DC30V 10A. Fig. 14.4 shows the circuit diagram of one channel of the relay board. All 4-channels have identical circuits. IN1, IN2, IN3, and IN4 are the active LOW inputs which means that a relay is activated when a logic LOW signal is applied to its input pin. A 1 K resistor and an LED are used

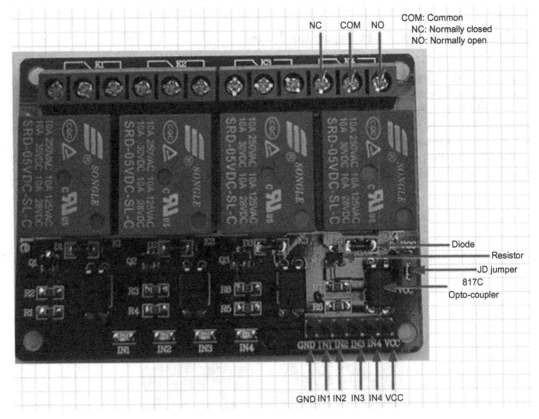

FIG. 14.3   4-Channel relay board.

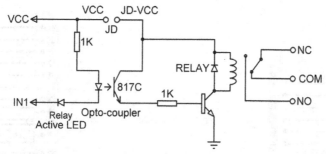

FIG. 14.4   Circuit diagram of one channel of the relay board.

in series with the input circuitry. With a 5 V power supply, assuming 2 V drop across the LED, each input circuit will draw about 3 mA current from the supply. Relay contacts are NC. Activating the relay changes the active contacts such that the common pin and NC pin become the two relay contacts and at the same time the LED at the input circuit of the relay board corresponding to the activated relay is turned ON. The VCC can be connected to either +3.3 V or to +5 V. Jumper JD is used to select the voltage for the relay. **Because the current drawn by a relay can be in excess of 80 mA, you must remove this jumper and connect an *external power supply* (e.g., +5 V) to pin JD-VCC.**

14.3.7 Circuit Diagram

Fig. 14.5 shows the circuit diagram of the project. IN1, IN2, IN3, and IN4 inputs of the relay board are connected to GPIO pins PC_0, PC_1, PC_2, and PC_3, respectively. Also, GND and +3.3 V pins of the development board are connected to GND and VCC pins of the relay board. You must make sure that jumper JD is removed from the board. Connect an external +5 V power supply to the JD-VCC pin of the relay board.

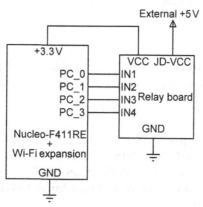

FIG. 14.5   Circuit diagram of the project.

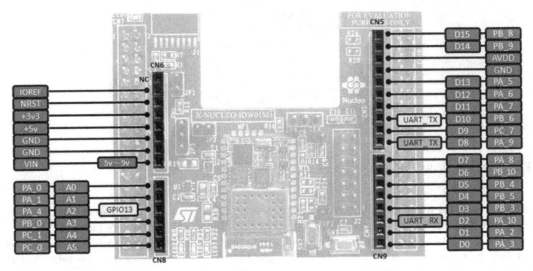

FIG. 14.6 Wi-Fi expansion board pin configuration.

The Wi-Fi expansion board pin configuration is shown in Fig. 14.6. Notice that UART1 pins PA_9 (TX) and PA_10 (RX) are used by the Wi-Fi expansion board.

14.3.8 The Construction

As shown in Fig. 14.7, the Wi-Fi expansion board was plugged on top of the Nucleo-F411RE development board and connections to the relay board were made using jumper wires.

14.3.9 The PDL

Fig. 14.8 shows the program PDL.

14.3.10 Program Listing

The program is named **WiFi**. The program uses the **X_NUCLEO_IDW01M1** library and the **NetworkSocketAPI** library. These libraries can be loaded to the program as follows:

- Create a new blank program and name it as **WiFi**.
- Open web site:
    https://os.mbed.com/components/X-NUCLEO-IDW01M1/
- Click on **Import Library**. Set **Target Path** to **WiFi** (see Fig.14.9).
- Right click on the **WiFi** project folder in **Program Workspace**.
- Click on **Import Library** and then click on **From Import Wizard**.
- Search for **Network Socket API**.
- Double click under name **NetworkSocketAPI** to load the library to your program.

FIG. 14.7   Connections to the relay board.

BEGIN
> Define development board to Wi-Fi board UART interface
> Define mobile device IP address
> Define Wi-Fi SSID name
> Define Wi-Fi password
> Configure all relays as outputs and assign them to PC_0 to _C_3
> Create a TCP socket
> Connect to Wi-Fi
> Connect to mobile device
> **DO FOREVER**
>> **IF** a data packet is received **THEN**
>>> **IF** the command is ONx **THEN**
>>>> Activate relay x
>>> **ELSE IF** command is OFFx **THEN**
>>>> De-activate relay x
>>> **ENDIF**
>> **ENDIF**
> **END**

END

FIG. 14.8   Program PDL.

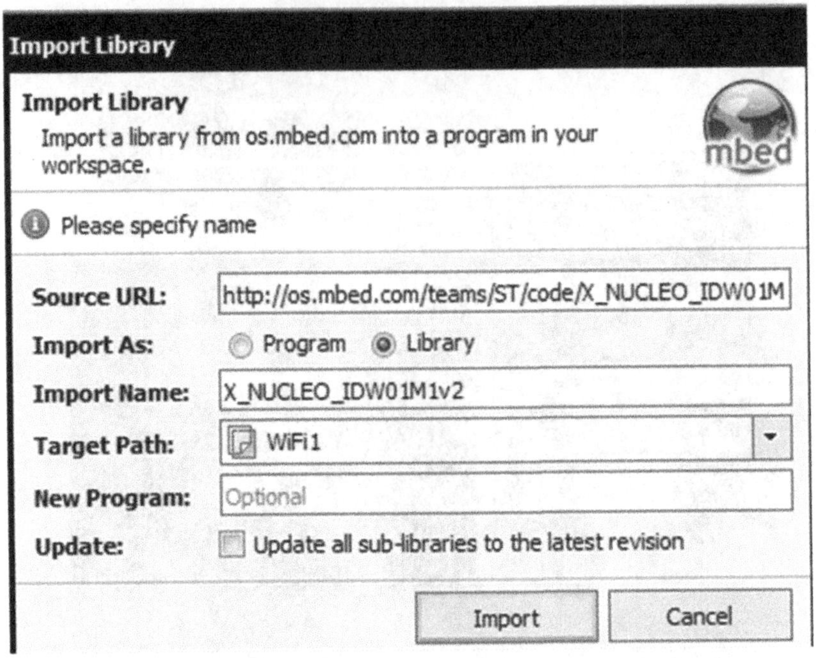

FIG. 14.9   Import the SpwfInterface library.

You should have the following folders and programs in your **Program Workspace**:

WiFi

> NetworkSocketAPI
> X_NUCLEO_IDW01M1v2
> > main.cpp
> > mbed

The program listing is shown in Fig. 14.10. At the beginning of the program header files **mbed.h**, **SpwfInterface.h**, and **TCPSocket.h** are included in the program. The UART interface between the Wi-Fi expansion board and the development board are then defined where PA_9 is the UART TX and PA_10 is UART RX. In this project the **IP** address of the mobile phone was **192.168.1.178** and this is assigned to **IPAddress**. The connections between the development board and the relay board are defined and PORTC pins PC_0 to PC_3 are assigned to variables **RELAY1** to **RELAY4**, respectively. Inside the main program the Wi-Fi **ssid** name and **password** are assigned to variables **ssid** and **password**, respectively. All the relays are deactivated at the beginning of the main program by setting the relay inputs to logic 1s. The program then connects to the mobile phone over a TCP socket with port number 5000 and waits in a loop to receive data (commands) from the mobile phone. The following actions are taken when data is received from the mobile phone:

```
/*************************************************************************
                  Wi-Fi EXPANDION BOARD RELAY CONTROL
                  ===================================

In this program the Nucleo Wi-Fi expansion board is plugged on top
of the NucleoOF411RE development board. PC_0 to PC_3 pins of the
development board are connected to a 4-channel relay board. The
relay contacts are controlled by sending commands from a mobile
phone (or a PC) over teh Wi-Fi link using the TCP protocol. Commands
ONx activates relay x. Similarly, command OFFx de-ctivates relay x.

Author: Dogan Ibrahim
Date  : September 2018
File  : WiFi
*************************************************************************/
#include "mbed.h"
#include "SpwfInterface.h"
#include "TCPSocket.h"

//
// Define the expansion board to development board UART connections
//
#define TX PA_9
#define RX PA_10
#define IPAddress "192.168.1.178"              // Mobile phone IP address
SpwfSAInterface wifi(TX, RX, false);

//
// Configure all relays as outputs
//
DigitalOut Relay1(PC_0);
DigitalOut Relay2(PC_1);
DigitalOut Relay3(PC_2);
DigitalOut Relay4(PC_3);


int main()
{
    int count;
    TCPSocket socket(&wifi);                // Create a TCP socket
    char Buffer[80];                        // Buffer to store data
    char *ssid = "BTHomeSpot-XNH";          // Wi-FI SSD name
    char *password = "49346abaeb";          // Wi-FI password

    Relay1 = 1;                             // De-activate Relay 1
    Relay2 = 1;                             // De-activate Relay 2
    Relay3 = 1;                             // De-activate Relay 3
    Relay4 = 1;                             // De-activate Relay 4

//
// Connect to the Wi-Fi
//
```

FIG. 14.10 Program listing.

(Continued)

```
    if(!wifi.connect(ssid, password, NSAPI_SECURITY_WPA2))

    {
        return -1;
    }

//
// Connect to the remote device (mobile phone) over TCP
//
    int ret = socket.connect(IPAddress, 5000);
    if(ret)return -1;

//
// Receive data packets over TCP and activate/de-activate the
// required relay. Variable count is greater than 0 if data is
// received over the network. The commands are ONx or OFFx. Thus
// for example Buffer[0] ='O' and Buffer[1]='N' or Buffer[0]='O'
// and Buffer[1]='F' and Buffer[2]='F'
//
    while(1)
    {
        count = socket.recv(Buffer, sizeof Buffer);
        if(count > 0)
        {
            if(Buffer[0] == 'O' && Buffer[1] == 'N')
            {
              if(Buffer[2] == '1')      Relay1 = 0;
              else if(Buffer[2] == '2')Relay2 = 0;
              else if(Buffer[2] == '3')Relay3 = 0;
              else if(Buffer[2] == '4')Relay4 = 0;
            }
            else if(Buffer[0] == 'O' && Buffer[1] == 'F' && Buffer[2] == 'F')
            {
              if(Buffer[3] == '1')      Relay1 = 1;
              else if(Buffer[3] == '2')Relay2 = 1;
              else if(Buffer[3] == '3')Relay3 = 1;
              else if(Buffer[3] == '4')Relay4 =1;
            }
        }
    }
}
```

FIG. 14.10, CONT'D

Command	Action
ON1	Activate Relay 1
ON2	Activate Relay 2
ON3	Activate Relay 3
ON4	Activate Relay 4
OFF1	Deactivate Relay 1
OFF2	Deactivate relay 2
OFF3	Deactivate Relay 3
OFF4	Deactivate Relay 4

Character array **Buffer** stores the data received from the mobile phone. Several **if-else** statements are used to determine the action to be taken depending on the received data (command). Notice that in this project the mobile device is the **Server** and the Wi-Fi expansion board is the **Client**.

14.3.11 Testing the Program

The program can easily be tested with a mobile phone. You should install a **TCP Server** apps on your Android mobile phone before starting the test. There are many freely available TCP apps in the **Play Store**. The one installed and used in this project is called the **Simple TCP Socket Tester** by *Armando J.G. Parra* (sabiocharro@gmail.com) as shown in Fig. 14.11.

The steps to test the program are as follows:

- Plug the Wi-Fi expansion board on top of your Nucleo-F411RE development board.
- Compile and download the program to your development board.
- Start the **Simple TCP Socket Tester** apps on your mobile phone and set the Port number to 5000, and click **START SERVER** so that the device starts to *listen for* incoming TCP connections.
- Press the **Reset** button on the development board and wait until a connection is made to the mobile device (you should see the notice: **Local IP: 192.168.1.178—Clients connected: 1** at the top of the screen, where 192.168.1.178 is the IP address of the mobile device). Now, activate relay 1 by sending the command **ON1** and click the **SEND** button as shown in Fig. 14.12. Relay 1 should now be activated on the relay board and at the same time the red LED corresponding to relay 1 should turn ON.

14.3.12 Suggestions for Additional Work

Modify the project so that the relays are controlled from a PC instead of a mobile phone.

14.4 PROJECT 2—ANDROID—NUCLEO BOARD COMMUNICATION USING THE ESP-01

14.4.1 Description

In this project we shall be using the tiny ESP-01 board to provide Wi-Fi capability to our development board. ESP-01 is the cheap alternative to the Nucleo Wi-Fi expansion board. Fig. 14.13 shows the ESP-01 module. This module has a few I/O ports and can either be used on its own as a standalone processor with Wi-Fi capability, or it can be used with a host processor to provide Wi-Fi capability to the host. In this project ESP-01 is used to provide Wi-Fi capability to our Nucleo-F411RE development board.

As in the previous project, a 4-channel relay board is connected to the development board and the relays are controlled by the commands sent by an Android mobile phone over the Wi-Fi link using the UDP protocol.

Simple TCP Socket Tester
Armando Jesús Gómez Parra

PEGI 3

FIG. 14.11   Simple TCP Socket Tester apps for Android.

14.4.2 Aim

The aim of this project is to show how the very low-cost ESP-01 module can be used with the Nucleo-F411RE development board to give Wi-Fi capability to the board. In addition, the project shows how a UDP program can be developed to communicate with a UDP apps on a mobile device.

FIG. 14.12   Activating relay 1.

FIG. 14.13   ESP-01 module.

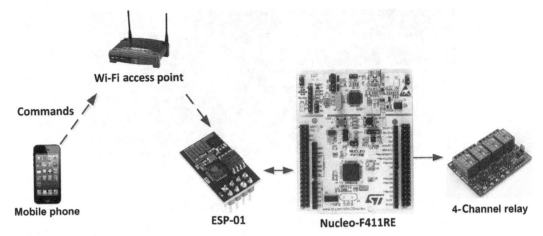

FIG. 14.14   Block diagram of the project.

14.4.3 Block Diagram

The block diagram of the project is shown in Fig. 14.14. The ESP-01 module is connected to the Nucleo-F411RE development board to provide it Wi-Fi capability. The relay board is connected and controlled from the development board.

14.4.4 Circuit Diagram

In this project an ESP-01-type Wi-Fi module is used. This module has an ESP8266 processor on board. The board is 8-pin and it operates with +3.3 V. The module is interfaced to a host processor via its TX and RX serial communication pins. The pin names are as follows:

VCC: +3.3 V power-supply pin
GND: Power-supply ground
GPIO0: I/O pin. This pin must be connected to +3.3 V for normal operation, and to GND for uploading firmware to the chip
GPIO2: General purpose I/O pin
RST: Reset pin. Must be connected to +3.3 V for normal operation
CH_PD: Enable pin. Must be connected to +3.3 V for normal operation
TX: Serial output pin
RX: Serial input pin

Fig. 14.15 shows the circuit diagram of the project. UART 1 is used on the development board to communicate with ESP-01. UART TX pin (PA_9) is connected to RX pin of ESP-01. Similarly, UART RX (PA_10) pin is connected to TX pin of ESP-01. ESP-01 is powered from the +3.3 V output voltage pin (pin 16, connector CN7) of the development board. The relay board is connected to the development board as in the previous project with jumper JD removed and external +5 V applied to its JD-VCC pin.

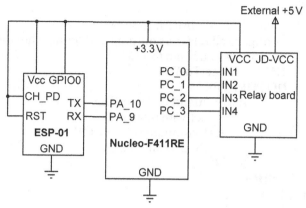

FIG. 14.15   Circuit diagram of the project.

14.4.5 The Construction

As shown in Fig. 14.16, the project was constructed on a breadboard and jumper wires were used to connect the ESP-01 to the development board. Also, the relay board was connected to the development board using jumper wires. Notice that the ESP-01 module is not breadboard compatible and an adaptor is necessary to plug it onto a breadboard.

14.4.6 The PDL

Fig. 14.17 shows the program PDL.

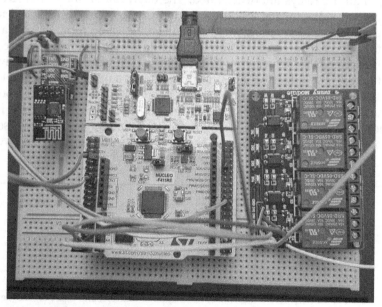

FIG. 14.16   Project constructed on a breadboard.

```
BEGIN/MAIN
        Define ESP-01 UART interface
        Configure all relays as outputs and de-activate them
        Set ESP-01 baud rate to 115200
        CALL ConnectToWiFi
        DO FOREVER
                IF command is ONx THEN
                        Activate relay x
                ELSE IF comman is OFFx THEN
                        De-activate relayx
                ENDIF
        ENDDO
END/MAIN


BEGIN/ConnectToWiFi
        Send commands to ESP-01 to connect to Wi-Fi
END/ConnectToWiFi
```

FIG. 14.17   Program PDL.

14.4.7 Program Listing

The program listing (program: **esp**) is shown in Fig. 14.18. At the beginning of the program Nucleo-F411RE development board UART1 TX and UART1 RX pins are assigned to **esp** pins RX and TX, respectively. Then, the GPIO ports connected to the relays are configured as digital outputs and all the relays are deactivated. The baud rate of the UART is set to 115200 which is the default value for the ESP-01. Function **ConnectToWiFi** is called to connect to the Wi-Fi. The remainder of the program runs in an endless loop formed using a **while** statement. Inside this loop data is received from the mobile device and the relay is controlled accordingly. Commands **ONx** activate relay **x**. Similarly, commands **OFFx** deactivates relay **x**. Function **strstr** looks for a substring in a string and returns a nonzero value if the substring is found. The reason for using the **strstr** function is because the data received from the mobile device is in the following format **+ID0,n: data** (e.g., **+ID0,3:ON1**) where 0 is the link ID and n is the number of characters received. Using the **strstr** function we can easily search for the strings ONx or OFFx. Notice that logic HIGH deactivates a relay and logic LOW activates it.

AT+RST	Reset ESP-01
AT+CWMODE	Set ESP-01 mode (here it is set to Station mode)
AT+CWJAP	Set Wi-Fi ssid name and password
AT+CPIMUX	Set connection mode (here it is set to multiple connection)
AT+CIFSR	Returns the IP address (not used here)
AT+CIPSTART	Set TCP or UDP connection mode, destination IP address, and port number (here, UDP is used with port number set to 5000. Destination IP address is set to "0.0.0.0" so that any device can connect and send data over Port 5000).

```
/*********************************************************************
                    ESP-01 BOARD RELAY CONTROL
                    ===========================

In this program an ESP-01 module is used to provide Wi-Fi functions
to the Nucleo-F411RE development board. Additionally, a 4-channel
relay board is connected to GPIO ports PC_0 to PC_3 of the
development board. Commands ONx activate relay x. Similarly, commands
OFFx de-activate relay x.

Author: Dogan Ibrahim
Date  : September 2018
File  : esp
*********************************************************************/
#include "mbed.h"

//
// Define the expansion board to development board UART connections
//
#define TX PA_9                                       // UART TX
#define RX PA_10                                      // UART RX
Serial esp(TX, RX);                                   // UART TX,RX

//
// Configure all relays as outputs and de-activate all relays
//
DigitalOut Relay1(PC_0, 1);
DigitalOut Relay2(PC_1, 1);
DigitalOut Relay3(PC_2, 1);
DigitalOut Relay4(PC_3, 1);

//
// This function connects to the Wi-Fi
//
void ConnectToWiFi()
{
    esp.printf("AT+RST\r\n");
    wait(2);

    esp.printf("AT+CWMODE=1\r\n");
    wait(3);

    esp.printf("AT+CWJAP=\"BTHomeSpot-XNH\",\"49346abaeb\"\r\n");
    wait(10);

    esp.printf("AT+CPIMUX=1\r\n");
    wait(3);

    esp.printf("AT+CIFSR\r\n");
    wait(3);

    esp.printf("AT+CIPSTART=\"UDP\",\"192.168.1.178\",0,5000,2\r\n");
    wait(3);
}

int main()
```

FIG. 14.18   Program listing.

(Continued)

```
{
    char Buffer[80];                    // Buffer to store data
    esp.baud(115200);                   // Set baud rate to 115200
    ConnectToWiFi();                    // Connect to Wi-Fi

// Receive data packets over UDP and activate/de-activate the
// required relay. Function strstr(a,"b") returns a value greater than
// 0 if string a contains sub-string "b". The command sent by the
// mobile devuce must be terminated with an Enter key (required for
// the scanf function). Buffer is cleared at the beginning of the loop
//
    while(1)
    {
        for(int k = 0; k < 5; k++)Buffer[k] = 0;

        esp.scanf("%s",&Buffer);
        if(strstr(Buffer, "ON1") > 0)        Relay1 = 0;
        else if(strstr(Buffer,"ON2") > 0)   Relay2 = 0;
        else if(strstr(Buffer, "ON3")> 0)    Relay3 = 0;
        else if(strstr(Buffer, "ON4") > 0)  Relay4 = 0;
        else if(strstr(Buffer, "OFF1") > 0)Relay1 = 1;
        else if(strstr(Buffer, "OFF2") > 0)Relay2 = 1;
        else if(strstr(Buffer, "OFF3") > 0)Relay3 = 1;
        else if(strstr(Buffer, "OFF4") > 0)Relay4 = 1;
    }
}
```

FIG. 14.18, CONT'D

Function **ConnectToWiFi** sends the following commands to the ESP-01 to connect to the Wi-Fi: Notice that small delays are used after each command. Command AT+CWJAP requires longer delay.

14.4.8 Testing the Program

The program can easily be tested with a mobile phone. You should install a **UDP Server** apps on your Android mobile phone before starting the test. There are many freely available UDP apps in the **Play Store**. The one installed and used in this project is called the **UDP Sender** by *hastarin* (hastarin@gmail.com) as shown in Fig. 14.19.

The steps to test the program are as follows:

- Construct the circuit.
- Compile and download the program to your development board.
- Start the **UDP Server** apps on your mobile phone and set the Port number to 5000 and the destination IP address (ESP-01 IP address, which was 192.168.1.160 in this project). Enter command **ON1** followed by the Enter key (required for the **scanf** function) on the mobile phone. Click Send to send the command to the development board. Relay 1 should now become active and the red LED corresponding to this relay will turn ON. Fig. 14.20 shows the command entered on the mobile phone.

FIG. 14.19   UDP Server apps for Android.

Notice that the IP address of the ESP-01 can be obtained by scanning all the devices on the local Wi-Fi router. For example, Android apps called **Wifi Watch** by *Riverdevs* (riverdevs@gmail.com) can be used to see the IP addresses of all the devices connected to the router. The ESP-01 is listed as shown in Fig. 14.21 (IP: 192.168.1.160).

14.5 PROJECT 3—ANDROID—NUCLEO BOARD COMMUNICATION USING BLUETOOTH

14.5.1 Description

In this project a Bluetooth module is connected to the Nucleo-F411RE development board. Additionally, two LEDs are connected to the development board. The LEDs are controlled

FIG. 14.20   Send command to activate relay 1.

(turned ON or OFF) from an Android mobile phone (or any other device with network access, e.g., PC or tablet) by sending commands through the Bluetooth interface. The LEDs are turned ON and OFF by sending commands ONx and OFFx where x is A for LEDA and B for LEDB.

14.5.2 Aim

The aim of this project is to show how the Nucleo-F411RE development board can communicate with external devices using Bluetooth communications.

14.5.3 Block Diagram

The block diagram of the project is shown in Fig. 14.22. The HC-06 Bluetooth module is used in this project. This is a serial module that can be controlled from the UART of the development board.

192.168.1.152

Hostname: Unknown
MAC: 10:f0:05:9f:ab:55
Manufacturer: Intel Corporate

192.168.1.160

Hostname: Unknown
MAC: a0:20:a6:14:c0:51
Manufacturer: Espressif Inc.

192.168.1.178

Hostname: HUAWEI_P20_Pro-a20f75d621
MAC: 02:00:00:00:00:00
Manufacturer: Unknown

FIG. 14.21 Finding the IP address of the ESP-01.

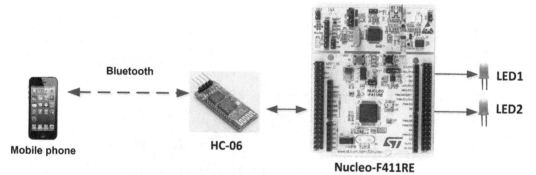

FIG. 14.22 Block diagram of the project.

14.5.4 Circuit Diagram

The HC-06 (see Fig. 14.23) is a four-pin serial Bluetooth module having the following features:

- +3.3 to +6 V operation
- 30 mA unpaired current (10 mA matched current)

FIG. 14.23   HC-06 Bluetooth module.

- built-in antenna
- default communication: 9600 baud, 8 data bits, no parity, 1 stop bit
- signal coverage 30 ft
- four-pin module

Fig. 14.24 shows the project circuit diagram. The TX and RX pins of HC-06 are connected to UART RX (PA_10) and UART TX (PA_9) pins of the development board. Two LEDs are connected to GPIO pins PC_0 and PC_1 with the names 1 and 2.

14.5.5 The Construction

As shown in Fig. 14.25, the project was constructed on a breadboard and connections made to the development board using jumper wires.

14.5.6 The PDL

Fig. 14.26 shows the program PDL.

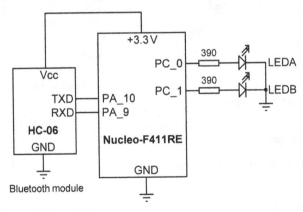

FIG. 14.24   Circuit diagram of the project.

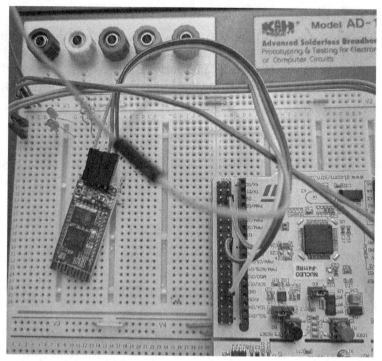

FIG. 14.25   Project constructed on a breadboard.

BEGIN
      Define HC-06 to development board interface
      Configure LEDA and LEDB as digital outputs
      Turn OFF LEDA and LEDB
      DO FOREVER
            Get a command
            IFcommand is ONA **THEN**
                  Turn ON LEDA
            ELSE IFcommand is ONB **THEN**
                  Turn ON LEDB
            ELSE IFcommand is OFFA **THEN**
                  Turn OFF LEDA
            ELSE IFcommand is OFFB **THEN**
                  Turn OFF LEDB
            ENDIF
      ENDDO
END

FIG. 14.26   Program PDL.

14.5.7 Program Listing

The program listing (program: **Bluetooth**) is shown in Fig. 14.27. At the beginning of the program the interface between the HC-06 Bluetooth module and the development board are defined and **LEDA** and **LEDB** are assigned to GPIO ports PC_0 and PC_1, respectively. These port pins are configured as digital outputs and the two LEDs are turned OFF at the beginning of the

```
/***********************************************************************
                        BLUETOOTH LED CONTROL
                        =====================

In this program a serial HC-06 Bluetooth module is connected to UART
pins PA_9 and PA_10. Additonally, two LEDs are connected to PORTC
pins PC_0 and PC_0. Commands sent from a Bluetooth device (e.g. a
mobile phone or a laptop) control the LEDs. Commands ONx turn ON
LEDs where x is A for LEDA and B for LEDB. Similarly, commands
OFFx turn OFF the LEDs where x is A fo LEDA and B for LEDB

Author: Dogan Ibrahim
Date   : September 2018
File   : Bluetooth
***********************************************************************/
#include "mbed.h"

#define TX PA_9                              // UART TX pin
#define RX PA_10                             // UART RX pin
Serial bluetooth(TX, RX);                    // Bluetooth interface

//
// Configure the LEDs as outputs and turn them OFF at beginning
//
DigitalOut LEDA(PC_0, 0);
DigitalOut LEDB(PC_1, 0);

int main()
{
    char Buffer[10];
//
// Receive commands and control the LEDs accordingly. The received
// commands are stored in character array Buffer.
//
    while(1)
    {
        bluetooth.scanf("%s", &Buffer);
        if(Buffer[0] == 'O' && Buffer[1] == 'N')
        {
            if(Buffer[2] == 'A')     LEDA = 1;
            else if(Buffer[2] == 'B')LEDB = 1;
        }
        else if(Buffer[0] == 'O' && Buffer[1] == 'F' && Buffer[2] == 'F')
        {
            if(Buffer[3] == 'A')     LEDA = 0;
            else if(Buffer[3] == 'B')LEDB = 0;
        }
    }
}
```

FIG. 14.27   Program listing.

program. The remainder of the program runs in an endless loop. Inside this loop data (commands) are received from the Bluetooth device using the **scanf** function and are stored in character array **Buffer**. The program then controls the LEDs based on these commands, for example, **ONA** turns LEDA ON, **ONB** turns LEDB ON, **OFFA** turns OFF LEDA, and **OFFB** turns OFF LEDB.

14.5.8 Testing the Program

The program can be tested by using an Android mobile phone to send commands through a Bluetooth communication interface. There are many freely available Bluetooth communication programs in the Play Store. The one chosen by the author was called the **Bluetooth Controller** by *mightyIT* (it@memighty.com) as shown in Fig. 14.28. You should

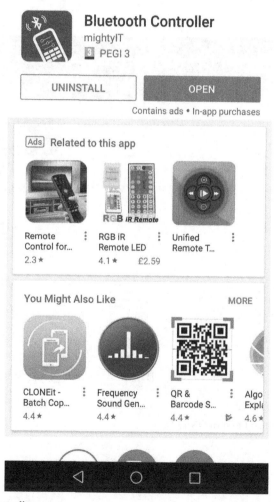

FIG. 14.28 Bluetooth Controller apps.

install this apps on your Android mobile phone so that you can send commands to the development board.

The steps to test the application are as follows:

- Construct the project.
- Compile and download the program to your Nucleo-F411 development board.
- Activate the **Bluetooth Controller** apps on your mobile phone.
- The apps will look for nearby Bluetooth devices. Click on **HC-06** when displayed on the mobile phone screen.
- You will now be asked to enter the password to pair the mobile phone with the development board. Enter the default password of **1234.** You should see a green color dot at the top right-hand side of the screen when a connection is made to the development board. Also, HC-06 with its address (e.g., HC-06 [98:D3:31:FB:5E:B6]) should be displayed at the top left side of the screen.
- To turn ON LEDA enter command **ONA** and click **Send ASCII**. You should see LEDA turning ON. Enter command **OFFA** to turn OFF LEDA. Fig. 14.29 shows an example screen.

FIG. 14.29   Example command to turn ON LEDA.

We can modify the program in Fig. 14.27 by sending a confirmation to the mobile phone when there is change in the LED status. The modified program listing (program: **BluetoothConfirm**) is shown in Fig. 14.30 where **printf** statements are used to send confirmation to the mobile device. For example, as shown in Fig. 14.31, the text LEDB is ON is displayed on the screen after the command **ONB** is sent.

```
/***********************************************************************
                    BLUETOOTH LED CONTROL
                    =====================

In this program a serial HC-06 Bluetooth module is connected to UART
pins PA_9 and PA_10. Additonally, two LEDs are connected to PORTC
pins PC_0 and PC_0. Commands sent from a Bluetooth device (e.g. a
mobile phone or a laptop) control the LEDs. Commands ONx turn ON
LEDs where x is A for LEDA and B for LEDB. Similarly, commands
OFFx turn OFF the LEDs where x is A fo LEDA and B for LEDB

In this version of the program, confirmation is sent back to the
mobile device when the status of an LED changes

Author: Dogan Ibrahim
Date  : September 2018
File  : BluetoothConfirm
***********************************************************************/
#include "mbed.h"

#define TX PA_9                                 // UART TX pin
#define RX PA_10                                // UART RX pin
Serial bluetooth(TX, RX);                       // Bluetooth interface

//
// Configure the LEDs as outputs and turn them OFF at beginning
//
DigitalOut LEDA(PC_0, 0);
DigitalOut LEDB(PC_1, 0);

int main()
{
    char Buffer[10];
//
// Receive commands and control the LEDs accordingly. The received
// commands are stored in character array Buffer.
//
```

FIG. 14.30   Modified program.

(Continued)

```
while(1)
{
    bluetooth.scanf("%s", &Buffer);
    if(Buffer[0] == 'O' && Buffer[1] == 'N')
    {
        if(Buffer[2] == 'A')
        {
            LEDA = 1;
            bluetooth.printf("LEDA is ON");
        }
        else if(Buffer[2] == 'B')
        {
            LEDB = 1;
            bluetooth.printf("LEDB is ON");
        }
    }
    else if(Buffer[0] == 'O' && Buffer[1] == 'F' && Buffer[2] == 'F')
    {
        if(Buffer[3] == 'A')
        {
            LEDA = 0;
            bluetooth.printf("LEDA is OFF");
        }
        else if(Buffer[3] == 'B')
        {
            LEDB = 0;
            bluetooth.printf("LEDB is OFF");
        }
    }
}
}
```

FIG. 14.30, CONT'D

14.6 SUMMARY

In this chapter we have learned about the following:

- Wi-Fi
- Nucleo Wi-Fi expansion board
- Mbed Wi-Fi functions
- 4-Channel relay board
- ESP-01
- Using ESP-01 in a project
- HC-06 Bluetooth
- Using Bluetooth in a project

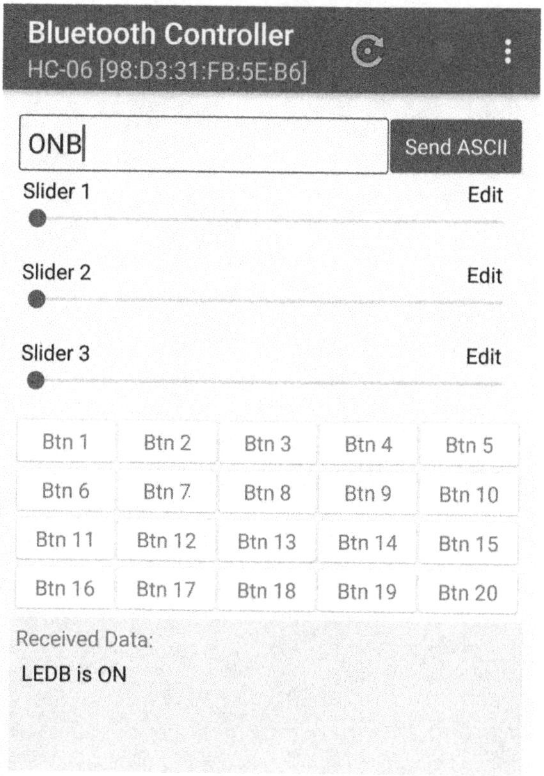

FIG. 14.31 Example display.

14.7 EXERCISES

1. Explain where and how the Nucleo Wi-Fi expansion board can be used.
2. A Nucleo Wi-Fi expansion board is connected to a Nucleo-F411RE development board. Write a program to display the IP address of the board on the PC screen.
3. A Nucleo-Wi-Fi expansion board and an LCD are connected to a Nucleo-F411RE development board. Write a program to display the IP address of the board on the LCD.
4. Explain how an Android mobile phone can communicate with the Nucleo-F411RE development board over a Wi-Fi link.
5. Explain the advantages and disadvantages of Bluetooth compared to Wi-Fi.
6. Explain how two devices can communicate over the Bluetooth interface.

Mbed RTOS Projects

15.1 OVERVIEW

Real-time operating system (RTOS) is very powerful extension to the Mbed operating system as it allows multiple tasks to run on the processor. There are many applications such as the Internet of Things (IOT) where it may be required to run multiple tasks, usually independent of each other on the same processor. In this chapter we shall be looking at the basic principles of RTOS in general and then develop several projects using the Mbed RTOS functions. In a RTOS the processor responds to external events very fast and in an orderly manner, switching between different tasks as governed by the scheduling algorithm used.

15.2 TASK SCHEDULING

Tasks can be defined to be small self-contained codes that usually run independent of each other in a program. For example in a multidigit 7-segment display application it is required to refresh the display frequently. This process can be implemented as a task independent of the other codes running in the program. Another example may be, suppose that it is required to flash an LED every second in a program. At the same time it may be required to check the status of a push-button switch and take appropriate actions when the button is pressed. In such applications we can either use interrupts to process different tasks, or use a multitasking approach if the programming language that we are using supports it. In multitasking approach the LED flashing code can run as an independent task and we can have another task to check the status of the button. Although the processor can execute only one task at any time, the scheduling algorithms used in multitasking operating systems switch between different tasks quickly so that it seems that different tasks execute at the same time.

Scheduling is the fundamental concept in multitasking systems. Basically, there are two scheduling algorithms: *nonpreemptive* and *preemptive*. It is important to know the difference between the two algorithms and this is described briefly in the following subsections.

15.2.1 Nonpreemptive Scheduling

Nonpreemptive scheduling is the simplest form of task scheduling in a multitasking system. Here, once a task is given the CPU, the CPU cannot be taken away from that task. It is up to the task to give away the CPU and this usually happens when the task completes its operations, or when the task is waiting for some external resources and thus cannot continue. Nonpreemptive scheduling is also called *Cooperative Scheduling*. Some characteristics of cooperative scheduling are:

- In cooperative scheduling short tasks have to wait for long tasks to complete before they can grab the CPU. In extreme cases short tasks can never be executed if the long tasks do not release the CPU.
- In cooperative scheduling there is no task priority and tasks that require immediate attention have to wait until other tasks release the CPU. This may not be desirable in many real-time applications where immediate attention of the CPU may be required, for example, when an important and highly risky external event occurs (e.g., very high temperature, alarm, etc.).
- It is important to make sure that a task does not block, for example, waiting for an external event to occur. If a task blocks then none of the other tasks can run in the system. A blocking task should release the CPU and wait for its turn to come again to check whether or not the blocking condition still exists.

Because of its nature, cooperative scheduling-based multitasking is not used in real-time systems where immediate attention of the CPU may be required. Fig. 15.1 shows an example cooperative scheduling of three tasks. Task1 takes the longest time and when it releases the CPU then Task2 starts. Task3 starts after Task2 completes its processing. CPU is given back to Task1 after Task3 releases it. Depending on the algorithm used, the context of a task may or may not be saved. Saving the context of a task enables the task to return and continue from the point where it released the CPU.

Perhaps the simplest way to implement cooperative scheduling in a program is to use a state diagram approach where the tasks can be selected using a **switch** statement as in the following skeleton code. In this example there are four tasks which are selected sequentially one after the other. Note that in this example the task context is not saved and tasks start from the beginning of their codes:

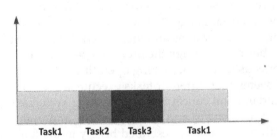

FIG. 15.1   Example cooperative scheduling with three tasks.

```
NextTask = 1;
while(1)
{
        switch(NextTask)
        {
                case 1:
                        Task1 code
                        NextTask = 2;
                        break;
                case 2:
                        Task2 code
                        NextTask = 3;
                        break;
                case 3:
                        Task3 code
                        NextTask = 4;
                        break;
                case 4:
                        Task4 code
                        NextTask = 1;
                        break;
        }
}
```

15.2.2 Preemptive Scheduling

In a preemptive scheduling once the CPU is given to a task it can be taken away, for example when a higher priority task wants the CPU. Preemptive scheduling is used in real-time systems where the tasks are usually configured with different priorities and time critical tasks are given higher priorities. A higher priority task can stop a lower priority one and grab and use the CPU until it releases it. In preemptive scheduling the task contexts are saved so that the tasks can resume their operations from the point they left when they are given back the CPU.

Preemptive scheduling is normally implemented in two different ways: using *Round Robin* (RR) scheduling, or using *interrupt-based* (IB) scheduling.

In RR scheduling all the tasks are given equal amount of CPU times and tasks do not have any priorities. When the CPU is to be given to another task, the context of the current task is saved and the next task is started. The task context is restored when the same task gets control of the CPU. RR scheduling has the advantage that all the tasks get equal amount of the CPU time. It is however not suitable in real-time systems since a time critical task cannot get hold of the CPU when it needs to. Also, a long task can be stopped before it completes its operations. Fig. 15.2 shows an example RR type scheduling with three tasks.

In IB scheduling tasks may be given different priorities and the task with the highest priority gets hold of the CPU. Tasks with same priorities are executed with RR type scheduling where they are all given equal amount of CPU time. IB scheduling is best suited to real-time systems where time critical tasks are given higher priorities. The disadvantages of IB

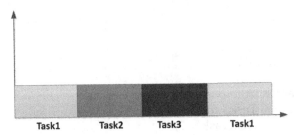

FIG. 15.2   Example Round Robin scheduling with three tasks.

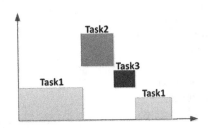

FIG. 15.3   Example IB scheduling with three tasks.

scheduling are that it is complex to implement, and also there is too much overhead in terms of context saving and restoring. Fig. 15.3 shows an example IB type scheduling with three tasks where Task1 has the lowest priority and Task2 has the highest priority.

15.3 Mbed RTOS IMPLEMENTATION

One of the nice features of Mbed is that it supports preemptive interrupt-based scheduling. Several tasks can be given different priorities, they can be scheduled to start, stop, communicate with other, and share resources. We shall be developing several multitasking projects in the next sections using various Mbed RTOS functions. *Note that the RTOS features described in this book refer to Mbed OS 2. Other OS versions may have different functions and features.*

15.4 PROJECT 1—DIFFERENT FLASHING A PAIR OF LEDs—USING Mbed THREAD

15.4.1 Description

In this project two LEDs (LEDA and LEDB) are connected to the Nucleo-F411RE development board. The project flashes LEDA every second, and LEDB every 0.5 s.

15.4.2 Aim

This is a multitasking project having two tasks. The aim of this project is to show how two tasks can be created using Mbed function Thread.

BEGIN/MAIN
       Configure PC_0 and PC_1 as digital outputs
       Assign LEDA and LEDB to PC_0 and PC_1 respectively
       Start thread LEDAControl
       Start thread LEDBControl
       Wait forever
END/MAIN

BEGIN/LEDAControl
       Flash LEDA every second
END/LEDAControl

BEGIN/LEDBControl
       Flash LEDB every second
END/LEDBControl

FIG. 15.4   Program PDL.

15.4.3 Circuit Diagram

LEDA and LEDB are connected to GPIO ports PC_0 and PC_1, respectively, through 390 ohm current limiting resistors.

15.4.4 The PDL

Fig. 15.4 shows the program PDL.

15.4.5 Mbed Thread

Function **Thread** allows defining and creating tasks in a program. **Main** in a program is a special case of a thread function that is started at system startup time and it has normal priority, known as **osPriorityNormal**.

A thread is basically a function in a program which normally runs continuously after it is started. Function **thread.start(name)** starts thread called **name**. Function **Thread::wait(n)** should be used to create **n** milliseconds of delay in a thread.

The program is called **MultiLED**. Before using the RTOS functions we have to load the RTOS library to our program. The steps are as follows:

- Create a blank program called **MultiLED**
- Go to the following web site:

  https://os.mbed.com/handbook/RTOS

- Click **Import Library** to import the RTOS library to your program space
- Select **Target Path** as **MultiLED**

Fig. 15.5 shows the program listing. At the beginning of the program, header files **mbed.h** and **rtos.h** are included in the program, **LEDA** and **LEDB** are configured as digital outputs are

```
/****************************************************************

                    DIFFERENT FLASHING LEDs
                    =======================

This is a simple multi-tasking application. In this program two
LEDs LEDA and LEDB are connected to PC_0 and PC_1 respectively.
LEDA flashes every second, while LEDB flashes every 500ms.

Author: Dogan Ibrahim
Date   : September 2018
File   : MultiLED
*****************************************************************/
#include "mbed.h"
#include "rtos.h"

DigitalOut LEDA(PC_0);                          // LEDA is output
DigitalOut LEDB(PC_1);                          // LEDB is output
Thread thread1, thread2;

//
// This is LEDA function
//
void LEDAControl()
{
    while(1)
    {
        LEDA = !LEDA;
        Thread::wait(1000);
    }
}


//
// This is LEDB function
//
void LEDBControl()
{
    while(1)
    {
        LEDB = !LEDB;
        Thread::wait(500);
    }
}


//
// This is the main program which starts the two threads
//
int main()
{
    thread1.start(LEDAControl);                 // Start LEDAControl
    thread2.start(LEDBControl);                 // Start LEDBControl

    while(1);                                   // Wait here forever
}
```

FIG. 15.5   Program listing.

assigned to PC_0 and PC_1, respectively. Two functions named **LEDAControl** and **LEDBControl** are created to flash the two LEDs. **LEDAControl** flashes LEDA every second, while **LEDBControl** flashes LEDB every 0.5 s. Inside the main program threads **LEDAControl** and **LEDBControl** are started. The main program loops forever, not doing anything useful. In actual fact, **main** is another thread and one of the LEDs could have been flashed inside the **main**. Because **main** is another thread, in this program there are actually three threads, two of them flashing the two LEDs and the third one just repeating itself in a loop and wasting CPU resources. We can place the **main** thread in a wait state so that it does not consume any CPU resources. This can be done by declaring a semaphore at the beginning of the program, such as:

Semaphore sema;

And then place the **main** thread in a forever wait state with the following statement:

sema.wait(osWaitForever);

We can also set the priority of the **main** to **idle** (see the following section) so that it does not consume any CPU resources. Another option is to set the thread to wait forever by the following statement:

Thread::wait(osWaitForever);

15.4.6 Mbed Thread Priorities

The default thread priority is **osPriorityNormal**. A thread can be set to one of the following priorities (Mbed OS2 only):

osPriotityIdle	lowest priority (−3)
osPriorityLow	low priority (−2)
osPriorityBelowNormal	below normal priority (−1)
osPriorityNormal	normal (default) priority (0)
osPriorityAboveNormal	above normal priority (+1)
osPriorityHigh	high priority (+2)
osPriorityRealtime	real-time (highest) priority (+3)

Table 15.1 gives a list of the important thread related functions with example usage (further details and complete list can be obtained from the web site: https://os.mbed.com/handbook/RTOS):

15.4.7 Thread States

A thread can be in any one of the following four states at any time:

RUNNING: this is the currently running thread (only one thread can be in this state at any time).

TABLE 15.1 Important Thread Functions

Function	Description	Example
join()	Wait for a thread to terminate	thread1.join()
terminate()	Terminate a thread	thread1.terminate()
set_priority()	Set thread priority	thread1.set_priority(osPriorityHigh)
get_priority()	Get thread priority	p = thread1.get_priority()
wait()	Wait specified milliseconds	Thread::wait(100)
gettid()	Get thread id of current thread	t = Thread::gettid()

READY: These are the threads which are ready to run. When the current thread relinquishes the CPU the next thread with the highest priority will run.
WAITING: the threads that are waiting for some events are in this state.
INACTIVE: threads that are terminated or are not started are in this state. These threads do not consume any CPU resources.

A WAITING thread can become READY or RUNNING when the event it is waiting for becomes available. A RUNNING thread becomes WAITING if it waits for an event. A RUNNING thread becomes READY if a higher priority thread becomes RAEDY. A thread becomes INACTIVE when it terminates.

15.4.8 Terminating Thread LEDAControl After 10 Flashes

In some applications we may want to terminate an active thread. We can for example modify the program given in Fig. 15.5 so that thread **LEDAControl** is terminated after 10 s. The **main** code for this new program is as follows the remainder of the program is same as in Fig. 15.5:

```
int main()
{
        thread1.start(LEDAControl);      // Start LEDAControl
        thread2.start(LEDBControl);      // Start LEDBControl

        Thread::wait(10000);             // Wait 10 seconds
        thread1.terminate();             // Terminate LEDAControl

        while(1);
}
```

15.4.9 Thread Callback

There are applications where we may want to pass parameters to a thread when the thread is started. This is done using the **callback** function. An example program (program: **ThreadCallback**) is shown in Fig. 15.6 where the flashing rate of the two LEDs are passed as arguments to the two functions where LEDA flashing rate is 250 ms and LEDB flashing rate is 500 ms. Note here that the addresses of the arguments must me passed to the functions and the functions use pointers to read the data.

```
/********************************************************************
                PASSING ARGUMENTS TO FUNCTIONS
                =================================

In this program two LEDs LEDA and LEDB are connected to PC_0 and
PC_1 as before. The LED flashing rates are passed as arguments
to the LEDs

Author: Dogan Ibrahim
Date   : September 2018
File   : ThreadCallback
*********************************************************************/
#include "mbed.h"
#include "rtos.h"

DigitalOut LEDA(PC_0);                    // LEDA is output
DigitalOut LEDB(PC_1);                    // LEDB is output
Thread thread1, thread2;

//
// This is LEDA function. The flashing rate is passed as an argument
//
void LEDAControl(int *tim)
{
    while(1)
    {
        LEDA = !LEDA;
        Thread::wait(*tim);
    }
}


//
// This is LEDB function. The flashing rat is passed as an argument
//
void LEDBControl(int *tim)
{
    while(1)
    {
        LEDB = !LEDB;
        Thread::wait(*tim);
    }
}


//
// This is the main program which starts the two threads with names
// LEDAControl and LEDBControl where both threads flash the LEDs.
// LEDAControl is terminates after 10 seconds
//
```

FIG. 15.6 Program listing with callback.

(Continued)

```
int main()
{
   int tim1 = 250, tim2 = 500;

   thread1.start(callback(LEDAControl, &tim1));      // Start LEDAControl
   thread2.start(callback(LEDBControl, &tim2));      // Start LEDBControl

   while(1);

}
```

FIG. 15.6, CONT'D

15.5 PROJECT 2—REFRESHING A 7-SEGMENT DISPLAY

15.5.1 Description

As we have seen in the project in Chapter 8.2, the digits of a multidigit 7-segment LED need to be refreshed at about every 5–10ms so that the human eye sees both digits to be ON at all times. The CPU has to spend all of its time to the refreshing process and as a result it cannot do any other tasks. In this project we use a multitasking approach where the refreshing process will be implemented in a separate task. In this project the display will count from 00 to 99 continuously with 1s delay between each count.

15.5.2 Aim

The aim of this project is to show how a multidigit 7-segment display can be refreshed in a multitasking environment.

15.5.3 Block Diagram

The block diagram of the project is as in Fig. 8.1.

15.5.4 Circuit Diagram

The circuit diagram of the project is as in Fig. 8.3 where the display segments are connected to PORT C pins of the development board, and an NPN transistor is used to control the display digits.

15.5.5 Program Listing

The program listing (program: **RTOS7Segment**) is shown in Fig. 15.7. The program is basically very similar to the one given in Fig. 8.5, except that here the display refreshing is done as a separate task. At the beginning of the program header files mbed.h and rtos.h are included in the program and the RTOS library is loaded to the program. The interface between the 7-segment display and the development board are then defined as in the program in Fig. 8.5 and global

```
/************************************************************************

                    2-DIGIT MULTIPLEXED LED REFRESHING
                    ===================================

In this program a 2-digit 7-Segment LED is connected to PORT C of the
Nucleo-F411RE development board. A multi-tasking approach is used where
the dislay is refeshed in a separate task to the main program. The
display counts from 00 to 99 continuously with one second delay between
each count

Author: Dogan Ibrahim
Date   : September 2018
File   : RTOS7Segment
************************************************************************/
#include "mbed.h"
#include "rtos.h"

PortOut Segments(PortC, 0xFF);
int LEDS[] = {0x3F,0x06,0x5B,0x4F,0x66,0x6D,0x7D,0x07,0x7F,0x6F};
Thread thread1, thread2;

DigitalOut MSDEnable(PC_8);
DigitalOut LSDEnable(PC_9);

#define Enable 1
#define Disable 0
int CNT;

//
// This is the display refreshing thread whoch runs as an
// independent task
//
void Refresh()
{
    int MSDValue, LSDValue;
    MSDEnable = Disable;
    LSDEnable = Disable;                // Disable LSD digit

    while(1)                            // Do forever
    {
        MSDValue = CNT / 10;            // MSD of the number
        LSDValue = CNT % 10;            // LSD of the number
        Segments = LEDS[MSDValue];      // Send to PORT C
        MSDEnable = Enable;             // Enable MSD digit
        Thread::wait(5);                // Wait 5ms

        MSDEnable = Disable;            // Disable MSD digit
        Segments = LEDS[LSDValue];      // Send to PORT C
        LSDEnable = Enable;             // Enable LSD digit
        Thread::wait(5);                // Wait 5ms
        LSDEnable = Disable;            // Disable LSD digit
    }
}
```

FIG. 15.7 Program listing.

(Continued)

```
//
// The main program increments CNT every second
//
int main()
{
    CNT = 0;                                    // CNT = 0
    thread1.start(Refresh);                     // Start thread Refresh

    while(1)                                    // Do forever
    {
        CNT++;                                  // Increment CNT
        if(CNT == 100)CNT = 0;                  // BAck to 0
        Thread::wait(1000);                     // Wait 1 second
    }
}
```

FIG. 15.7,   CONT'D

variable CNT is defined. Inside the main program variable CNT is cleared to 0 and thread Refresh is started. This thread refreshes the display by enabling every digit for 10 ms. The main program increments CNT every second and when it reaches 100 it resets back to 0.

15.6 Mbed TASK SYNCHRONIZATION—MUTEX, SEMAPHORE, AND SIGNALS

Mbed provides a number of functions for synchronizing threads when multiple threads need to access common global variables. Mutexes and semaphores are used for this purpose. Signals are used to synchronize threads to the occurrence of certain events. For example, an event can be forced to wait for an event to occur and as soon as the event occurs the event can be signaled to continue. In this section we shall be looking at the mutexes, semaphores, and signals in some detail.

15.6.1 Mutexes

Mutexes (or mutual exclusion objects) are programming objects that allow multiple program threads to share the same resource, such as data or file. A mutex is given a unique name when it is created. The resource is shared in locked by a thread. Once locked, other threads cannot access this resource until it is released by the thread that locked it. Therefore, a mutex is like a lock of a shared resource. A deadlock situation arises if a resource required by other threads are locked and never released by a mutex. A mutex is owned by the thread that uses it to lock the resource and any other thread cannot unlock this mutex. It is important to realize that a mutex locks part of a thread and also any data inside this thread. Mbed functions **lock()** and **unlock()** are used to lock and unlock a mutex, respectively.

Fig. 15.8 shows an example program (program: **Mutex**). In this example two threads are created called **task1** and **task2** which both print messages before and after they run. Without locking the following messages are displayed on the screen:

Before 1
Before 2
After 1
After 2

```
#include "mbed.h"
#include "rtos.h"

Thread thread1, thread2;
Serial pc(USBTX, USBRX);

void Count(int i)
{
    pc.printf("Before %d\n\r",i);      // Display Before i
    pc.printf("After %d\n\r",i);       // Display After i
}

void task1()
{
    Count(1);
}

void task2()
{
    Count(2);
}

int main()
{
    thread1.start(task1);              // Start task1
    thread2.start(task2);              // Start task2
    thread1.join();                    // Wait until 1 finished
    thread2.join();                    // Wait until 2 finished
}
```

FIG. 15.8   Example program without mutex.

In Fig. 15.9 (program: **MutexEx**) the same program is given where a mutex is used to lock the shared **printf** resource. This modified program displays the following messages on the screen which is what is expected normally:

Before 1
After 1
Before 2
After 2

15.6.2 Semaphores

A semaphore is simply a nonnegative integer which increments and decrements and controls access to a shared resource. A semaphore is initially set to a count equivalent to the number of free resources. A semaphore's value must be positive to allow access to the shared resource. The semaphore count is decremented by one when a thread uses the semaphore. Similarly, the count is incremented by one when the thread releases the semaphore. A zero count does not allow access to the shared resource. Semaphores with only one count are similar to mutexes. Such semaphores are also known as Binary Semaphores. An analogy of semaphores is the usage of a printer (shared resource), for example, three users sending print requests all at the same

```
#include "mbed.h"
#include "rtos.h"

Thread thread1, thread2;
Serial pc(USBTX, USBRX);
Mutex shr;                              // Create a mutex

void Count(int i)
{
    shr.lock();                         // Lock the resource
    pc.printf("Before %d\n\r",i);       // Display Before i
    pc.printf("After %d\n\r",i);        // Display After i
    shr.unlock();                       // Unlock the resource
}

void task1()
{
    Count(1);
}

void task2()
{
    Count(2);
}

int main()
{
    thread1.start(task1);               // Start task1
    thread2.start(task2);               // Start task2
    thread1.join();                     // Wait until 1 finished
    thread2.join();                     // Wait until 2 finished
}
```

FIG. 15.9   Program using a mutex.

time, if all the jobs are to start in parallel at the same time then one user's output will overlap with another and the outputs will mix up. We can protect this process using semaphored so that the printer resource is blocked when one process is running and then unblocked when it is finished. This process is then repeated for each user so that the jobs do not overlap. Mbed **wait()** and **release()** functions are used to lock and unlock a semaphore.

An example program (program: **sema**) using a semaphore is given in Fig. 15.10. In this program four threads are created where they display the messages **First, Second, Third,** and **Fourth** on the screen. By using a semaphore with a count of 2 we restrict access at any time to the **printf** function so that only two messages are displayed. The other two messages are displayed after a delay of 2s, that is, the display is as follows:

First
Second
<2 seconds delay>
Third
Fourth
<2 seconds delay>
First
Second
....
....

```
#include "mbed.h"
#include "rtos.h"

Thread thread1, thread2, thread3, thread4;
Semaphore sema(2);                                // Create a semaphore
Serial pc(USBTX, USBRX);

void task(const char *txt)
{
    while(1)
    {
        sema.wait();                              // Wait on semaphore
        pc.printf("%s\n\r", (char *)txt);
        wait(2);
        sema.release();                           // Release semaphore
    }
}

int main()
{
    thread1.start(callback(task, "First"));       // Start thread1
    thread2.start(callback(task, "Second"));      // Start thread2
    thread3.start(callback(task, "Third"));       // Start thread3
    thread4.start(callback(task, "Fourth"));      // Start thread4
}
```

FIG. 15.10   Semaphore example.

15.6.3 Signals

Signals are also called event flags and they are used for thread synchronization. A thread can be forced to wait for a signal to be set by another thread before it can continue. Mbed function **signal_wait()** forces the issuing thread to wait until the specified signal is set. Function **signal_set()** sets a signal. We can have up to 32 signal flags (or event flags) per thread. Fig. 15.11 shows an example where one task generates some data and another one reads this data. The reading task waits for an event flag to indicate that data are available before it continues to read it.

An example program (program: **signal**) using signals is shown in Fig. 15.12. This program uses the on-board LED and the on-board button. A thread called **Flash** is created which waits for event flag 1 to be set by calling function **signal_wait(0x1)**. Event flag 1 is set when the button is pressed by calling function **signal_set(0x1).** After this point thread **Flash** becomes active and flashes the LED every second. Remember that you have to load the RTOS library to your program before it can be compiled. As we have seen in earlier projects, the on-board LED and button are named as **LED1** and **BUTTON1**, respectively.

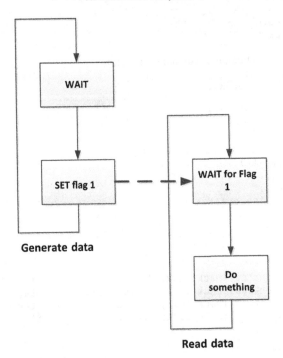

FIG. 15.11   Signals example.

```
#include "mbed.h"
#include "rtos.h"

Thread thread1;

DigitalOut led(LED1);
DigitalIn button(BUTTON1);

void Flash()
{
    Thread::signal_wait(0x1);          // Wait for event flag 1
    while(1)                           // Do forever
    {
        led = !led;                    // Flash teh LED
        Thread::wait(1000);            // Wait 1 second
    }
}

int main()
{
    thread1.start(Flash);              // Start thread1
    while(button == 1);                // Wait for button to eb pressed
    thread1.signal_set(0x1);           // Set eventflag 1

    while(1);                          // Wait forever
}
```

FIG. 15.12   Example program using signals.

15.7 PROJECT 3—CAR PARK CONTROLLER—USING SIGNALS

15.7.1 Description

This is a simple car park controller project. It is assumed that the car park has a capacity of 100 cars. Entry and exit gate barriers are available at the car park entry and exit points, respectively. The gate output signals are normally at logic HIGH and they go to logic LOW when a gate opens to let a car pass through. The controller counts the number of cars inside the car park and displays messages on the screen to let the drivers know how many spaces are available inside the car park (if any). The following information is displayed on the screen:

SPACES: nnnn

Initially the card park is assumed to be closed (e.g., at night) and message **CLOSED** is displayed. The car park becomes operational when a button called **StartButton** is pressed. When the car park is full or when it is closed the entry barrier is locked and does not open to let any cars into the car park.

15.7.2 Aim

The aim of this project is to show how a car park controller can be designed using a multitasking approach.

15.7.3 Block Diagram

The block diagram of the project is shown Fig. 15.13. It is assumed that barriers with switches are used at the entry and exit points of the car park. Normally the outputs of these switches are held at logic HIGH and they become LOW when the barriers are lifted.

15.7.4 Circuit Diagram

The circuit diagram of the project is shown in Fig. 15.14. Entry and exit switches are named as **EntrySwitch** and **ExitSwitch**, respectively and they are connected to GPIO pins PC_0 and PC_1 as shown in the figure. In this project two push-button switches are used to simulate the car park barriers. A PC is used to display the car park information.

15.7.5 The PDL

The PDL of the program is shown in Fig. 15.15.

15.7.6 Program Listing

The program listing (program: **CarPark**) is shown in Fig. 15.16. At the beginning of the program, header files **mbed.h** and **rtos.h** are included in the program, and the serial PC interface is defined. **EntrySwitch**, **ExitSwitch**, and **StartButton** are assigned to PC_0, PC_1, and PC_3,

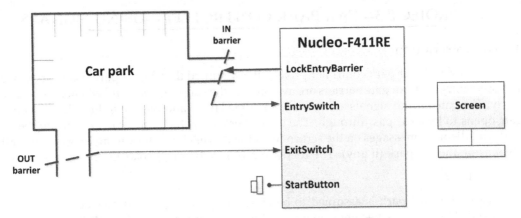

FIG. 15.13   Block diagram of the project.

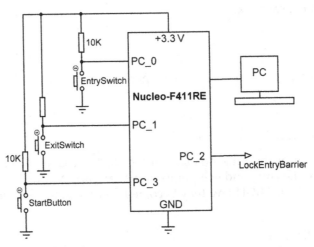

FIG. 15.14   Circuit diagram of the project.

respectively. **LockEntryBarrier** output when set locks the entry barrier so that it does not open and it is assigned to PC_2. The car park capacity is set to 100 cars. Inside the main program message **CLOSED** is displayed since the car park is closed at the beginning of the program. Threads **CarPark** and **Display** are started. These threads wait for event flags 1 and 2 to be set, respectively, before they can continue. These flags are set when the **StartButton** is pressed (i.e., **StartButton** is at logic LOW). Setting the flags starts the two threads. Thread **Display** displays the spaces available in the car park at screen coordinate (0, 9). Thread **CarPark** increases or decreases the space count depending on whether the cars are entering or leaving the car park. When a car enters the car park variable **Spaces** is decremented by one. Similarly, when a car leaves the car park variable **Spaces** is incremented by one. The program makes sure that the space count is not above the car park capacity or below zero (this should never happen in practice).

Note that this project could have been implemented without using multitasking, but here multitasking is used for demonstration purposes.

BEGIN/MAIN
      Lock the entry barrier
      Define car park capacity
      Display message CLOSED
      Start thread Display
      Start thread CarPark
      Wait for StartButton to be pressed
      Set event flag 1
      Set event flag 2
      Wait forever
END/MAIN

BEGIN/Display
      Wait for event flag 2
      DO FOREVER
            Display available spaces
      ENDDO
END/Display

BEGIN/CarPArk
      Qait for event flag 1
      DO FOREVER
            IF no spaces available **THEN**
                  Lock the entry barrier
            ELSE
                  Unlock the entry barrier
            ENDIF
            IF car entered **THEN**
                  Decrement the space count
                  IF space count is < 0 **THEN**
                        Space count = 0
                  ENDIF
            ELSE IF car left **THEN**
                  Increase the space count
                  IF space count > car park capacity **THEN**
                        Space count = ca park capacity
                  ENDIF
            ENDIF
      ENDDO
END/CarPark

FIG. 15.15   Program PDL.

```
/*********************************************************************

                    CAR PARK CONTROLLER
                    ===================

This is a car park controller program. The program keeps track of the
cars entering and leaving the car park. Available spaces is displayed
on the PC screen. Variable Spaces is decremented by one when a car
enters the car park. This variable is increased by one when a car
leaves teh car park.

The program is based on mult-tasking where thread Display displays
teh available space, and thread CarPArk caculates teh available spaces

Author: Dogan Ibrahim
Date  : September 2018
File  : CarPark
*********************************************************************/
#include "mbed.h"
#include "rtos.h"

Serial pc(USBTX, USBRX);                       // Serial PC interface

DigitalIn EntrySwitch(PC_0);                   // Entry switch
DigitalIn ExitSwitch(PC_1);                    // Exit switch
DigitalOut LockEntryBarrier(PC_2);
DigitalIn StartButton(PC_3);

Thread thread1, thread2;                       // Create threads
#define CarParkCapacity 100                    // Car park capacity
int Spaces;

//
// Clear the screen
//
void clrscr()
{
    char clrscr[] = {0x1B, '[', '2' , 'J',0};
    pc.printf(clrscr);
}

//
// Home the cursor
//
void homescr()
{
    char homescr[] = {0x1B, '[' , 'H' , 0};
    pc.printf(homescr);
}

//
// Goto specified line and column
//
```

FIG. 15.16   Program listing.

(Continued)

```
void gotoscr(int line, int column)
{
    char scr[] = {0x1B, '[', 0x00, ';' ,0x00, 'H', 0};
    scr[2] = line;
    scr[4] = column;
    pc.printf(scr);
}


//
// This is the CarPark thread. This thread calculates the
// available spaces based on the entry and exit barrier
// switches. The thread initially waits on an event flag
//
void CarPark()
{
    Thread::signal_wait(0x1);                // Wait for flag 1
    while(1)                                 // Do forever
    {
        if(Spaces == 0)                      // If no spaces...
            LockEntryBarrier = 1;            // Lock the barrier
        else                                 // Else...
            LockEntryBarrier = 0;            // Unlock the barrier

        if(EntrySwitch == 0)                 // If a car enters
        {
            Spaces--;                        // Decrement space count
            if(Spaces < 0)Spaces = 0;        // If no spaces
            while(EntrySwitch == 0);         // Wait until barrier down
        }

        if(ExitSwitch == 0)                  // If a car leaves
        {
            Spaces++;                        // Increment space count
            if(Spaces > CarParkCapacity)Spaces = CarParkCapacity;
            while(ExitSwitch == 0);
        }
    }
}


//
// This thread displays the car park status. Initially the thread
// waits for an event flag
//
void Display()
{
    Thread::signal_wait(0x2);                // Wait for flag 2
    homescr();                               // Home cursor
    pc.printf("SPACES = ");                  // Display Spaces =

    while(1)                                 // Do forever
    {
        gotoscr('0', '9');                   // Goto 0,9
        pc.printf("%d ", Spaces);            // Display spaces
        Thread::wait(250);                   // Wait a bit
    }
```

FIG. 15.6, CONT'D

(Continued)

```
    }


    //
    // This is the main program. The prgram locks the barrier to start
    // with and displays the message CLOSED. When the StartButton is
    // pressed flags 1 and 2 are set and therfore threads CarPark and
    // Display start running
    //
    int main()
    {
        Spaces = CarParkCapacity;           // At the beginning
        LockEntryBarrier = 1;               // Lock entry barrier (at night)
        clrscr();                           // Clear screen
        homescr();                          // Home cursor
        pc.printf("CLOSED");                // Display CLOSED

        thread1.start(CarPark);             // Start thread CarPark
        thread2.start(Display);             // Start thread Display

        while(StartButton == 1);            // Wait to start processing
                                            // StartButton is pressed...
        thread1.signal_set(0x1);            // Set flag 1
        thread2.signal_set(0x2);            // Set flag 2

        while(1);                           // Wait here forever
    }
```

FIG. 15.6,   CONT'D

FIG. 15.17   Typical display from the program.

A typical display from the program is shown in Fig. 15.17.

15.8 Mbed QUEUE AND MEMORYPOOL

Queues allow the user to queue pointers to data from producer threads to consumer threads. The Mbed queue functions are **queue.put()** to put into the queue and **queue.get()** to get from a queue. Fig. 15.18 shows a basic queue operation. A queue is created using the keyword **Queue**.

MemoryPool is used to manage fixed-size memory pools. The Mbed memory pool function **alloc()** is used to allocate a fixed amount of memory to the thread. Function **free()** returns the allocated memory block.

An example combined queue and memory pool program (program: **queue**) is shown in Fig. 15.19. In this program a structure called **msg** is created with two integer variables **no1**

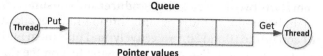

FIG. 15.18 Basic queue operation.

```
#include "mbed.h"
#include "rtos.h"
Thread thread1, thread2;
Serial pc(USBTX, USBRX);

typedef struct                          // Create a structure
{
    int no1;                            // First element
    int no2;                            // Second element
}msg;

MemoryPool<msg, 16> mpool;              // Create a memory pool
Queue<msg,16> queue;                    // Create a queue

//
// Generate data and put into queue
//
void producer()
{
    msg *message = mpool.alloc();
    message->no1 = 10;                  // No1 = 10
    message->no2 = 20;                  // No2 = 20
    queue.put(message);                 // Put into queue
}

//
// Get dada from the queue and display
//
void consumer()
{
    osEvent evt = queue.get();
    if(evt.status == osEventMessage)
    {
        msg *message = (msg*)evt.value.p;
        pc.printf("No1=%d No2=%d\n\r", message->no1, message->no2);
    }
}

//
// Create two threads: producer and consumer
//
int main()
{
    thread1.start(producer);            // Start thread1
    thread2.start(consumer);            // Start thread2

    while(1);

}
```

FIG. 15.19 Queue and MemoryPool example.

and **no2**. Main program starts two threads called **producer** and **consumer**. A memory pool and a queue are defined at the beginning of the program. Thread **producer** allocates a memory pool and sets variables **no1** and **no2** to 10 and 20, respectively, and puts them into the queue. Thread **consumer** reads from the queue and displays the following text on the PC screen:

no1 = 10 no2 = 20

Here, structure **osEvent** points to the actual data

15.9 Mbed MAIL

Mail is a very useful feature similar to a queue that is used to send and receive messages between threads, with the added advantage that it can allocate memory pool. Mbed Mail function **alloc()** allocates memory, **put()** puts a mail in the queue, **get()** gets a mail from the queue, and **free()** returns the allocated block.

An example program (program: **mail**) using Mail is shown in Fig. 15.20. In this program a structure is created as in the previous example with two integer numbers. A mail is created with the name **mailbox**. Thread producer allocates a memory pool, stores 5 and 20 in variables **no1** and **no2,** respectively. The data are then put into the mail using statement **mailbox.put(mail)**. The producer thread then waits for 2s before freeing the memory pool. Thread consumer gets the mail and then displays the contents of **no1** and **no2** on the PC screen. Note that **osEvent** is a pointer to the data.

15.10 Mbed RTOS TIMER

The RTOS timer can be used to create and control one-shot as well as periodic timer functions. A timer can be started, restarted, or stopped. Function **stop()** stops a timer, function **start(ms)** starts (or restarts) a timer where the argument is the timer period in milliseconds.

An example use of an RTOS timer program (program: **rtostmr**) is shown in Fig. 15.21. In this example an RTOS Timer is created with the name timer. This timer is then started with a period of 500ms. After 5s the timer is stopped.

15.11 SUMMARY

Multitasking is an important concept in the development of complex real-time systems. In this chapter we have learned the following:

- Multitasking scheduling algorithms
- Mbed RTOS
- Mbed thread, queue, signal, mutex, semaphore, mail, and RTOS timer
- Example projects using multitasking features of Mbed

```
#include "mbed.h"
#include "rtos.h"
Thread thread1, thread2;
Serial pc(USBTX, USBRX);

typedef struct                          // Create a structure
{
    int no1;                            // First element
    int no2;                            // Second element
}mymail;

Mail<mymail, 16> mailbox;               // Create a mail

//
// Generate data and put into mail
//
void producer()
{
    mymail *mail = mailbox.alloc();     // Allocate memory
    mail->no1 = 5;                      // No1 = 5
    mail->no2 = 20;                     // No2 = 20
    mailbox.put(mail);                  // Put into mail
    Thread::wait(2000);                 // Wait 2 seconds
    mailbox.free(mail);                 // Free memory
}

//
// Get dada from the mail and display it
//
void consumer()
{
    osEvent evt = mailbox.get();        // Get data
    if(evt.status == osEventMail)
    {
        mymail *mail = (mymail*)evt.value.p;
        pc.printf("No1=%d No2=%d\n\r", mail->no1, mail->no2);
    }
}

//
// Create two threads: producer and consumer
//
int main()
{
  thread1.start(producer);
  thread2.start(consumer);

  while(1);
}
```

FIG. 15.20 Example program using Mail.

```
#include "mbed.h"
#include "rtos.h"

DigitalOut led(LED1);

void Flash()
{
    led = !led;                      // Flash the LED
}

int main()
{
    RtosTimer timer(&Flash);         // Create a timer
    timer.start(500);                // Start the timer
    wait(5);                         // Wait 5 seconds
    timer.stop();                    // Stop the timer

    Thread::wait(osWaitForever);     // Wait forever

}
```

FIG. 15.21   RTOS Timer example.

15.12 EXERCISES

1. Explain what you understand from the term multitasking
2. Why is multitasking important in real-time systems?
3. Explain what a thread is?
4. Explain the advantages and disadvantages of the various scheduling algorithms
5. Three LEDs are connected to a Nucleo-F411RE development board. Write a program so the LEDs flash in the following rate: LED1 every second, LED2 every 0.5 s, LED3 every 100 ms.
6. Explain what a mutex is and where it can be used
7. Explain the differences between a mutex and a semaphore
8. Explain where a mail can be used in a multitasking system

Internet of Things (IoT)

16.1 OVERVIEW

Internet of Things (IoT) is currently one of the rapidly expanding uses of embedded processors in smart monitoring and control applications. In this chapter we shall be briefly looking at the principles of the IoT and give the complete design of a simple home IoT system.

16.2 INTERNET OF THINGS (IoT)

The IoT is a new concept in intelligent automation, monitoring, and control. It is the network of home appliances, vehicles, and any other physical devices connected together with sensors, actuators, and displays, which enables these devices to connect and exchange data. It is estimated that in 2017 there were around 9 billion IoT devices in the world and this number is estimated to grow over 30 billion by 2020.

The word "things" in IoT usually refers to devices that have unique identifiers, connected to the Internet to exchange information with each other in real time. Such devices have sensors and/or actuators that can be used to collect data about their environments and to monitor and control these environments as required. The collected data can be processed locally, or alternatively, it can be sent to centralized servers or the cloud for remote storage and processing.

Some IoT home applications include wireless or Internet connected smart appliances that can be turned on and off remotely using, say, a mobile phone or a tablet. For example, a small device can be used to collect data about temperature, humidity, and atmospheric pressure inside the house. This data can then be accessed at any time and from anywhere, provided there is Internet connectivity. Smart refrigerators can keep track of items stored and place orders automatically through the Internet with little or no interaction from their owners. Smart televisions can learn their owners' watching habits and inform them when a new show is available. Smart home lighting systems can turn on and off automatically and can change light intensities to adapt to the environment. For example, the light intensity can be reduced automatically in the day to save energy. Tea or coffee makers can turn on automatically when the owner wakes up. Smart smoke detectors can raise alarms in a friendly human voice, describing where the fire is and what actions to take. Smart car parks can inform drivers looking for a space to park and inform them the availability of the closest space. Smart roads can send messages to drivers to inform them about the road and weather conditions. Emergency services can use IoT technology to save lives and improve the environment. For example, gas leakage details can be sent automatically to emergency gas service departments to prevent any explosions or deaths. Forest fires can be detected by early warning IoT systems. There are also many applications of the IoT in the retail industry. For example, purchasing habits of customers can be stored by IoT systems which then inform them of special offers and discounts. IoT systems can be used in airports to inform passengers of flight delays before they leave their homes. Flight tickets and hotels can be booked automatically with simple voice commands. Wearable IoT devices can be used to monitor the health of the users, such as the blood pressure, temperature, heart beats, etc. Health centers can automatically be informed of patients' locations if a serious condition such as a heart attack is detected. There are many more examples of the use of IoT in present day.

16.2.1 IoT Architecture

There is no standard IoT architecture. An IoT system can make use of any technology to read sensors and send data to users. We can basically define two possible IoT architectures as far as the processor architecture is concerned: *distributed processors* and *common processor.*

Fig. 16.1 shows a typical distributed processor based IoT architecture. Here, each IoT device sends its data to local processors using a communication technology, such as Wi-Fi, Bluetooth, or simple RF. The actuators of the IoT devices are also connected to the same processors. These local processors then communicate with a central processor which can, for example, be a PC. The PC can send the data to a cloud so that the IoT devices can be monitored and controlled remotely from anywhere in the world and at any time of the day. Users can access the IoT devices through the Internet using their mobile phones, laptops, or tablets.

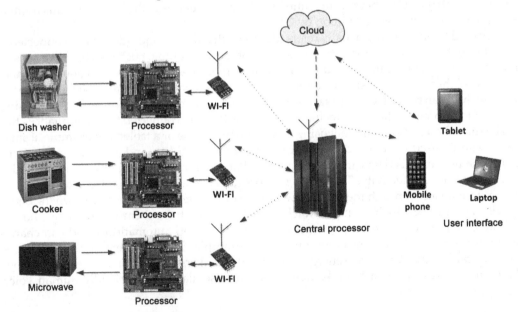

FIG. 16.1   IoT using distributed processors.

There can be several variations of the basic distributed processors architecture. In *shared distributed processors* architecture shown in Fig. 16.2, a group of IoT devices can share a local processor. This architecture has the advantages that it is simpler and also the cost is less since the local processor count is reduced. The disadvantage of this architecture is that some IoT devices may be located far away from the shared processors and it may not be possible to make direct connections. In this architecture the IoT devices can be connected directly to the local processors with wires and only one wireless communication device can be used.

Fig. 16.3 shows the IoT common processor architecture. Here, all the sensors as well as the actuators are connected to a local common processor using wires. The common processor can

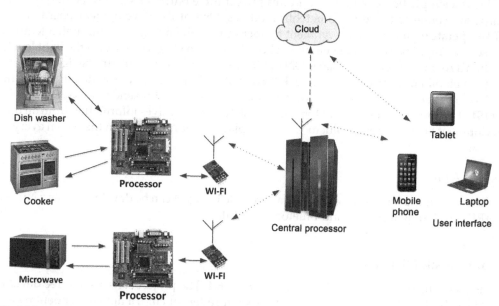

FIG. 16.2   IoT using shared distributed processors.

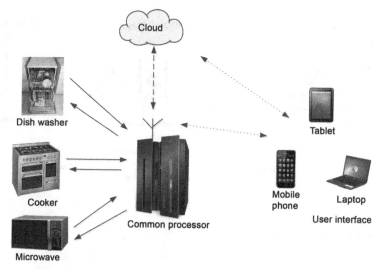

FIG. 16.3   IoT using a common processor.

then send the collected data to a cloud so that users can access the system from anywhere on Earth and at any time. The users can also access the system locally through the common processor.

16.3 PROJECT 1—HOME IoT PROJECT

16.3.1 Description

This is a simple home IoT project. In this project three Nucleo-F411RE-type development boards are connected to three Bluetooth modules. One of the development boards (node: **IoTTemperature**) measures the ambient temperature in the living room through a temperature sensor chip. The second development board (node: **IoTLight**) measures the light level in children's room and sends messages ON or OFF to informs the user about the light state. The third development board (called node: **IoTWindow**) is connected to a switch in the kitchen window and it checks whether or not the window is open and it sends messages **OPEN** or **CLOSED** to inform the user about the window state. A PC receives information from all three development boards through Bluetooth and displays this information on the screen every 5 s.

16.3.2 Aim

The aim of this project is to show how a simple IoT project can be developed using Nucleo-F411RE development boards and Bluetooth modules.

16.3.3 Block Diagram

The block diagram of the project is shown Fig. 16.4. The project consists of three development boards, three Bluetooth modules, a temperature sensor chip, a light dependent resistor (LDR), and a switch.

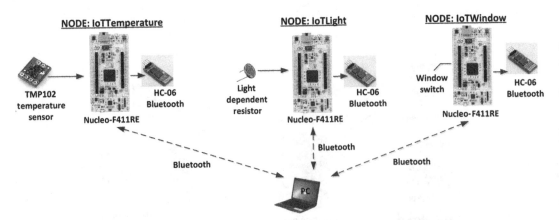

FIG. 16.4   Block diagram of the project.

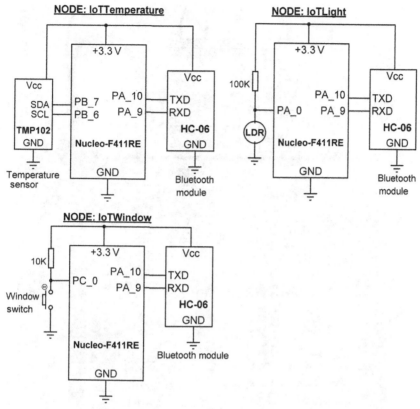

FIG. 16.5 Circuit diagram of the project.

16.3.4 Circuit Diagram

The circuit diagram of the project is shown in Fig. 16.5. Temperature sensor chip TMP102 is connected to I^{2}C port of node **IoTTemperature**. A LDR is connected to analog input PA_0 of node **IoTLight**. The LDR resistance is low at daylight and it becomes very high at dark. By measuring the voltage across the LDR we can tell whether the lights are ON or OFF in childrens' room. A switch mounted on the kitchen window is connected to digital port pin PC_0 of node **IoTWindow**. Normally the state of this switch is at logic HIGH to indicate that the window is open, and it goes to logic LOW when the window is closed. Bluetooth modules are connected to UART ports of all three development boards so that they can communicate with the PC.

16.3.5 The Construction

Fig. 16.6 shows the project built using three Nucleo-F411RE development boards.

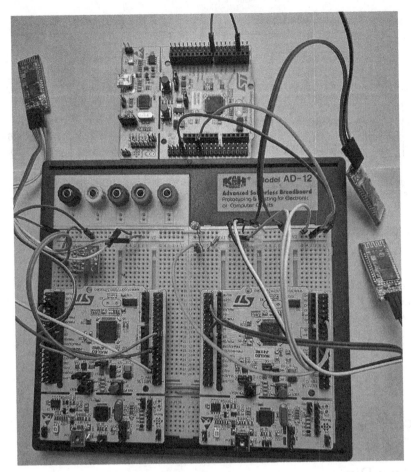

FIG. 16.6   The project built using three development boards.

16.3.6 The PDL

Fig. 16.7 shows the program PDL of each node.

16.3.7 Program Listing

There are three programs in this project, one for each node. Fig. 16.8 shows the program (program: **IoTTemperature**) for node **IoTTemperature**. At the beginning of this program the Bluetooth port is defined where TX is PA_9 and RX is PA_10 respectively. The I^2C interface between the development board and the TMP102 temperature sensor chip is defined. As explained in Chapter 11, the program reads the ambient temperature. The temperature is converted into degrees Centigrade and is sent through the Bluetooth module every 15 s using statement **bluetooth.printf** as a floating point data in the format %5.2f, that is, in the form **nn.nn**.

BEGIN/IoTTemperature
> Define I2C interface between the development board and TMP102
> Define the Bluetooth interface
> **DO FOREVER**
>> Read the temperature
>> Convert into Degrees Centigrade
>> Send heading through Bluetooth
>> Send the temperature through Bluetooth
>> Wait 15 seconds
> **ENDDO**

END/IoTTemperature

BEGIN/IoTLight
> Define the Bluetooth interface
> **DO FOREVER**
>> Read the LDR voltage
>> Convert into lux
>> Find out whether or not the lights are ON
>> Send heading through Bluetooth
>> **IF** light level is very low **THEN**
>>> Send LIGHTS are OFF message through Bluetooth
>> **ELSE**
>>> Send LIGHTS are ON message through Bluetooth
>> **ENDIF**
>> Wait 15 seconds
> **ENDDO**

END/IoTLight

BEGIN/IoTWindow
> Define Bluetooth interface
> **DO FOREVER**
>> Find out if the switch is pressed
>> Send heading through Bluetooth
>> **IF** switch is closed **THEN**
>>> Send WINDOW is CLOSED message through Bluetooth
>> **ELSE**
>>> Send WINDOW is OPEN message through Bluetooth
>> **ENDIF**
>> Wait 15 seconds
> **ENDDO**

END/IoTWindow

FIG. 16.7   Program PDL.

```
/**********************************************************************

                        IoT TEMPERATURE SENSOR
                        =========================


In this project a TMP102 type temperature sensor module is connected to
I2C1 port of the Nucleo-F411RE development board. The program reads and
sends the ambient temperature to a PC over the Bluetooth communication
interface. The data is sent every 15 seconds.

Author: Dogan Ibrahim
Date   : October 2018
File   : IoTTemperture
**********************************************************************/
#include "mbed.h"
I2C TMP102(PB_7, PB_6);                              // I2C SDA, SCL

#define TX PA_9                                      // Bluetooth port
#define RX PA_10                                     // Bluetooth port

Serial bluetooth(TX, RX);                            // Bluetooth TX, RX

const int TMP102Address = 0x90;                      // TMP102 address
char ConfigRegister[3];                              // Config register
char TemperatureRegister[2];                         // Temperature Register
float Temperature;

//
// Clear the screen
//
void clrscr()
{
    char clrscr[] = {0x1B, '[', '2' , 'J',0};
    bluetooth.printf(clrscr);
}


//
// Home the cursor
//
void homescr()
{
    char homescr[] = {0x1B, '[' , 'H' , 0};
    bluetooth.printf(homescr);
}



//
// This function configures the TMP102 after power up or reset
//
void ConfigureTMP102()
{
    ConfigRegister[0] = 0x01;                        // Point to Config Register
    ConfigRegister[1] = 0x60;                        // Upper byte
    ConfigRegister[2] = 0xA0;                        // Lower byte
    TMP102.write(TMP102Address, ConfigRegister, 3);  // Write 3 bytes
```

FIG. 16.8   Program IoTTemperature.

(Continued)

```
    }


int main()
{
    unsigned short M;
    char L;

    ConfigureTMP102();                                   // Configure TMP102
    ConfigRegister[0] = 0x00;                            // Point to Temp Register
    TMP102.write(TMP102Address, ConfigRegister, 1); // Write 1 byte
//
// Read the ambient temperature and send it to the PC over the Bluetooth
// interface every 15 seconds
//
    while(1)
    {
        TMP102.read(TMP102Address, TemperatureRegister, 2);
        M = TemperatureRegister[0] << 4;
        L = TemperatureRegister[1] >> 4;
        M = M + L;
        Temperature = 0.0625 * M;
        clrscr();
        homescr();
        bluetooth.printf("Living Room Temperature\n\r");
        bluetooth.printf("----------------------\n\r");
        bluetooth.printf("\n\r");
        bluetooth.printf("Temperature = %5.2fC\n\r", Temperature);
        wait(15.0);
    }
}
```

FIG. 16.8,   CONT'D

Fig. 16.9 shows the program (program: **IoTLight**) for node **IoTLight**. At the beginning of this program the Bluetooth port is defined where TX is PA_9 and RX is PA_10, respectively. PA_0 is used as the analog port to read voltage across the LDR. The light level is converted into lux. It is found by experiment that when the room is dark the light level is below 50. Therefore, the message **Lights are OFF** is displayed if the room is dark, otherwise message **Lights are ON** is displayed through the Bluetooth module every 15 s.

Fig. 16.10 shows the program (program: **IoTWindow**) for node **IoTWindow**. At the beginning of this program the Bluetooth port is defined where TX is PA_9 and RX is PA_10, respectively. PC_0 is used as digital input port to read the state of the window switch. If the switch is at logic HIGH then the message **Window is OPEN** is displayed, otherwise message **Window is CLOSED** is displayed through the Bluetooth module every 15 s.

```
/*************************************************************************

                        IoT LIGHT LEVEL
                        ===============

In this project an LDR is used to sense the light level in a room. The
program finds out whether the room lights are ON of OFF and sends
messages to the PC over the Bluetooth communication interface to
indicate the status of the lights

Author: Dogan Ibrahim
Date  : October 2018
File  : IoTLight
*************************************************************************/
#include "mbed.h"

#define TX PA_9
#define RX PA_10
Serial bluetooth(TX, RX);                       // Bluetooth interface

AnalogIn LDR(PA_0);                             // Analog port

//
// Clear the screen
//
void clrscr()
{
    char clrscr[] = {0x1B, '[', '2' , 'J',0};
    bluetooth.printf(clrscr);
}

//
// Home the cursor
//
void homescr()
{
    char homescr[] = {0x1B, '[' , 'H' , 0};
    bluetooth.printf(homescr);
}


int main()
{
    double mV, R, lux;

//
```

FIG. 16.9   Program IoTLight.

(Continued)

```
// Read the output voltage of the LDR, then calculate the resistance
// of it and finally determine whether or not the lights are ON or OFF
//
    while(1)                                        // Do forever
    {
        mV = 3300.0f * LDR.read();                  // In mV
        R = 10.0 * mV / (3300.0 - mV);              // LDR resistance
        lux = 2.17 - 1.28 * log10(R);               //
        lux = pow(10.0, lux);                       // Light level (Lux)
        clrscr();                                   // Clear screen
        homescr();
        bluetooth.printf("Light in Children's Room\n\r");
        bluetooth.printf("------------------------\n\r");
        bluetooth.printf("\n\r");
        if(lux < 50.0)
            bluetooth.printf("Lights are OFF\n\r");
        else
            bluetooth.printf("Lights are ON\n\r");
        wait(15.0);                                 // Wait 15 seconds
    }
}
```

FIG. 16.9, CONT'D

16.3.8 Pairing the Bluetooth Modules With the PC

It is necessary to pair the Bluetooth modules with the PC before they can exchange data with the PC. The steps for this are given in the following (these steps are for Window 7.0 operating system, but similar steps should work for other versions of Windows). It is assumed in the following that the Bluetooth modules are attached to all the development boards:

- Un-plug all development boards from the power supply
- Enable Bluetooth on your laptop or desktop
- Plug-in one of the development boards to the power supply
- Click the Bluetooth icon on your laptop and select **Show Bluetooth Devices**
- Click **Add a device**. You should see **HC-06** listed as shown in Fig. 16.11
- Click on **HC-06** and then click Next
- Click on **Pair Without Using Code**. You should see a message that the device has been added successfully as shown in Fig. 16.12.
- Click **Close**
- Repeat the above process for the other two development boards. At the end, you should see three HC-06 devices in the **Show Bluetooth Devices** list.

16.3.9 The PC Program

The **TeraTerm** terminal emulation program is used in this project for the PC. This emulation program has the option that it can connect to the PC Bluetooth ports and receive or send data through these ports. **TeraTerm** can be downloaded free of charge from several sources on the Internet.

```
/*********************************************************************

                    IoT WINDOW STATE CHECKING
                    =========================

This program checks the state of a switch connected to a window
and sends a message to the PC over the Bluetooth communication
interface to notify whether the window is open or closed.

Author: Dogan Ibrahim
Date   : October 2018
File   : IoTWindow
**********************************************************************/
#include "mbed.h"

#define TX PA_9
#define RX PA_10
Serial bluetooth(TX, RX);                        // Bluetooth interface

DigitalIn window(PC_0);                          // Window switch input

//
// Clear the screen
//
void clrscr()
{
    char clrscr[] = {0x1B, '[', '2' , 'J',0};
    bluetooth.printf(clrscr);
}


//
// Home the cursor
//
void homescr()
{
    char homescr[] = {0x1B, '[' , 'H' , 0};
    bluetooth.printf(homescr);
}


//
// If teh switch state is logic HIGH then the window is assumed to
// be open, otherwise it is assumed to be closed. The state of the
// window is checked every 15 seconds
//
int main()
{
```

FIG. 16.10   Program IoTWindow.

(Continued)

```
while(1)
{
    clrscr();
    homescr();
    bluetooth.printf("Kitched Window State\n\r");
    bluetooth.printf("-------------------\n\r");
    bluetooth.printf("\n\r");
    if(window == 1)
        bluetooth.printf("Window is OPEN\n\r");
    else
        bluetooth.printf("Window is CLOSED\n\r");
    wait(15.0);
}
}
```

FIG. 16.10, CONT'D

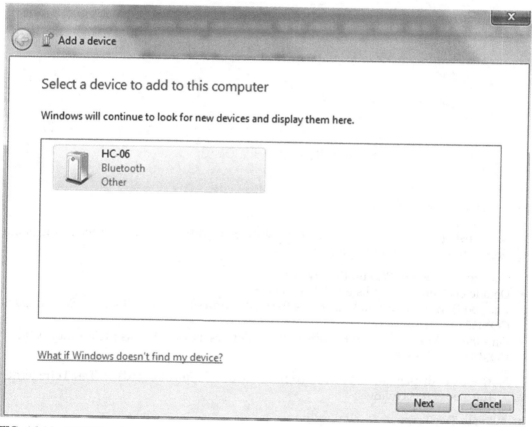

FIG. 16.11   HC-06 listed.

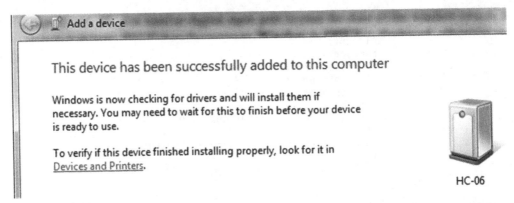

FIG. 16.12   Device added successfully.

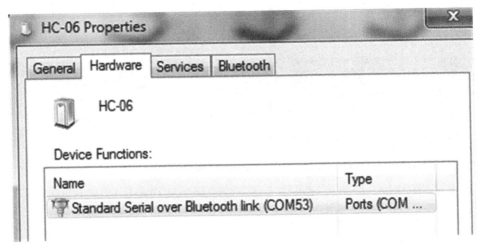

FIG. 16.13   Find the COM address.

Before using the **TeraTerm** we have to know the port addresses of the connected Bluetooth devices. These can be found as follows:

- Click to open **Show Bluetooth Devices**
- Double click on the first listed **HC-06** device
- Click on **Hardware** and make a note of the COM address. In Fig. 16.13 the COM address is **COM53**.
- Find the COM addresses of the other HC-06 devices as well. In this project they were **COM55** and **COM57**

Now we can display the data received from the development boards on **TeraTerm** terminals. The steps are as follows:

- Activate **TeraTerm**
- Click **Serial**
- Click Port and select **COM53: Standard Serial Over Bluetooth** as shown in Fig. 16.14

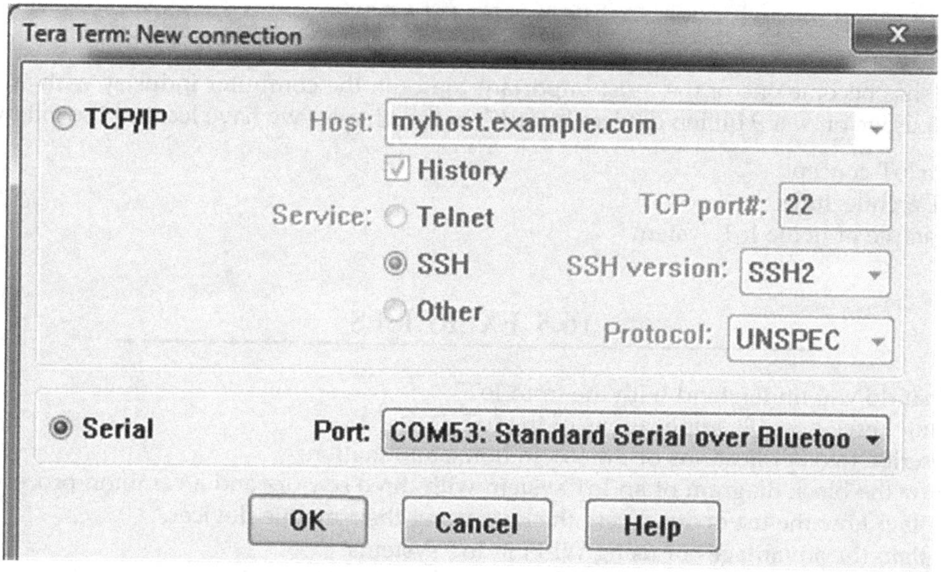

FIG. 16.14   Select port COM53.

FIG. 16.15   All the TeraTerm windows side by side.

- You should see the message **A Bluetooth device is trying to connect** at the bottom right-hand side of the screen. Click on this message and then enter the default pairing password as **1234** and click **Next**
- Click **Close**
- Activate two other **TeraTerm** sessions and enter the other two COM addresses
- You should now have three **TeraTerm** windows open

You can see all the **TeraTerm** windows by clicking on **Window** and then for example clicking **Side by Side** in any of the **TeraTerm** windows. Fig. 16.15 shows the three windows displaying the living room temperature, children's room lighting, and state of the kitchen window.

16.4 SUMMARY

IoT has become one of the most important topics in the computer industry with an estimated usage of over 9 billion devices by 2017. In this chapter we have learned the following:

- The IoT concept
- IoT architectures
- Example of home IoT system

16.5 EXERCISES

1. What do you understand with the term IoT?
2. Name three possible applications of the IoT.
3. Describe two applications of the IoT in home automation.
4. Draw the block diagram of an IoT system with three sensors and a common processor. Explain how the users can access the data using their mobile devices.
5. Explain the advantages of using Wi-Fi in IoT systems.
6. What are the disadvantages of using Bluetooth in IoT systems?

CHAPTER

17

STM32 Nucleo Expansion Boards

17.1 OVERVIEW

The expansion boards plug on top of the Nucleo development boards and provide additional peripheral components and modules such as sensors, actuators, displays, and so on. These are professional quality boards that can be used during project development.

The expansion boards are equipped with Arduino or Morpho compatible connectors so that they can be plugged on top of the Nucleo development boards. The expansion boards are supported by STM32-based software modules (e.g., TrueSTUDIO, Keil MDK-ARM, System Workbench, STM32CubeMX) as well as by Mbed, thus making it very easy to develop complex projects in relatively short times. The combination of the STM32 Nucleo board and expansion boards is a scalable approach for application development, prototyping, or for product evaluation.

In this chapter we shall be looking at the features of some of the commonly used expansion boards and also develop some projects using some of the popular expansion boards. Further information on the expansion boards can be obtained from the following ST website:

http://www.st.com/en/evaluation-tools/stm32-nucleo-expansion-boards.html?querycriteria=productId=SC1971

The pictures of the expansion boards in this chapter and the information contained in the chapter are a Copyright of STMicroelectronics and have been used here with their written permission: ©**STMicroelectronics. Used with permission**.

17.2 HIGH-POWER STEPPER MOTOR BOARD (X-NUCLEO-IHM03A1)

This is a stepper motor driver expansion board (Fig. 17.1) designed for driving high-power bipolar stepper motors, and is based on powerSTEP01. The board is compatible with the Arduino UNO R3 connector. The basic features of this expansion board are as follows:

- voltage range from 10.5 to 85 V
- phase current up to 10 A rms

FIG. 17.1   X-NUCLEO-IHM03A1 expansion board.

- power OK and fault LEDs
- up to 1/128 microstepping
- programmable speed profile and overcurrent protection
- sensorless stall detection
- adjustable output slew rate
- overtemperature protection

17.3 TWO-AXIS STEPPER MOTOR BOARD (X-NUCLEO-IHM02A1)

This is a two-axis stepper motor driver expansion board (Fig. 17.2) based on two L6470 motor driver chips. The board is equipped with Arduino UNO R3 connectors. The board can drive one or two stepper motors through USART interface. The basic features of this expansion board are as follows:

- 8–45 V operating voltage
- 7 A peak (3 A rms) output current for each motor driver
- USART communication with a PC
- SPI interface that may be connected in a daisy chain configuration
- five LEDs to indicate power, busy, and fault conditions

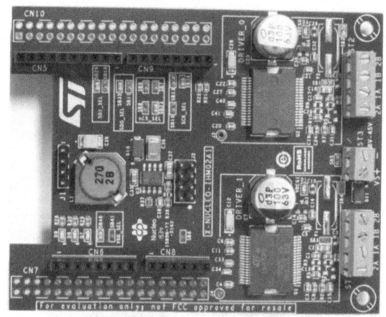

FIG. 17.2   X-NUCLEO-IHM02A1 expansion board.

17.4 LOW-VOLTAGE THREE-PHASE BRUSHLESS DC MOTOR BOARD (X-NUCLEO-IHM11M1)

This is a low-voltage three-phase brushless DC motor driver expansion board (Fig. 17.3) based on the STSPIN230. The board is compatible with the Arduino UNO R3 connector. The board is designed for six-step and FOC algorithms. The basic features of this expansion board are as follows:

- 1.8–10 V operation
- up to 1.3 A rms current
- overcurrent and short-circuit protection
- thermal shutdown
- current control with adjustable off time
- hall/encoder sensor connector and circuit
- potentiometer for speed regulation

17.5 MOTION MEMS AND ENVIRONMENTAL SENSOR EXPANSION BOARD (X-NUCLEO-IKS01A2)

This is a motion MEMS and environmental sensor expansion board (Fig. 17.4), equipped with three-dimensional (3D) accelerometer, 3D gyroscope, 3D magnetometer, humidity and temperature sensor, and ambient atmospheric pressure sensor. The board is interfaced via the

FIG. 17.3   X-NUCLEO-IHM11M1 expansion board.

I^2C pin and is compatible with the Arduino UNO R3 connector. The basic features of this expansion board are as follows:

- LSM6DSL MEMS 3D accelerometer ($\pm2/\pm4/\pm8/\pm16$ g) and 3D gyroscope
- LSM303AGR MEMS 3D accelerometer ($\pm2/\pm4/\pm8/\pm16$ g) and MEMS 3D magnetometer (±50 G)
- LPS22HB MEMS pressure sensor (260–1260 hPa)
- HTS221 capacitive relative humidity and temperature sensor
- DIL24 socket for additional MEMS sensors (e.g., UV index)
- I^2C interface

This board is described in more detail since a project is given in this section which displays the current humidity, temperature, and atmospheric pressure. The X-NUCLEO-IKS01A2 expansion board is equipped with jumpers which allow separate current consumption measurement of each sensor. An additional sensor can be connected as an adapter board to J1 DIL24 socket.

The board functional block diagram is shown in Fig. 17.5. There are a number of jumpers that can be used to configure the various modules on the board. All the modules on the board are configured to operate as I^2C slaves by default (JP7: 1, 2, 3, 4 connected, also JP8: 1, 2, 3, 4 connected)

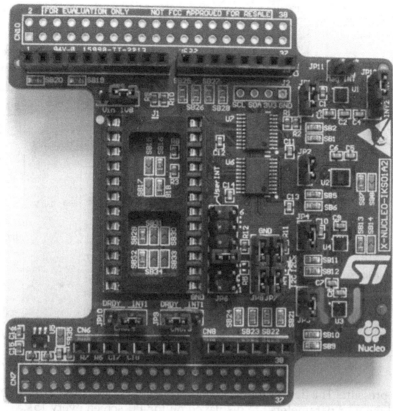

FIG. 17.4   X-NUCLEO-IKS01A2 expansion board.

and they all share the same bus. The module I²C default addresses are shown in Fig. 17.5 and are repeated below for convenience:

Module	I²C Address	Module Description
LSM6DSL	0xD6	Accelerometer + Gyroscope
LSM303AGR	0x32	Accelerometer
LSM303AGR	0x3C	Magnetometer
HTS221	0xBE	Humidity + Temperature
LPS221HB	0xBA	Pressure

17.6 PROJECT 1—MEASURING AND DISPLAYING THE HUMIDITY, TEMPERATURE, ATMOSPHERIC PRESSURE, AND DEW POINT USING THE X-NUCLEO-IKS01A2 EXPANSION BOARD

17.6.1 Description

In this project an X-NUCLEO-IKS01A2 expansion board is plugged on the Nucleo-F411RE development board. The project measures the ambient humidity, temperature, and the

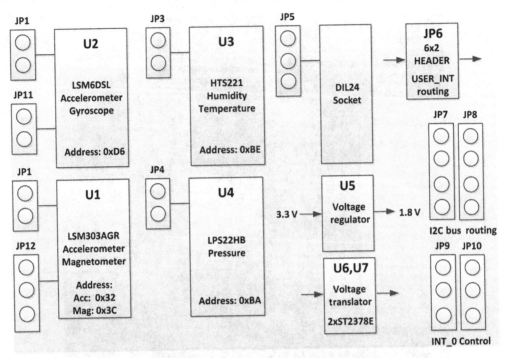

FIG. 17.5 Functional block diagram.

atmospheric pressure. The dew point is calculated from knowledge of the humidity and temperature. All the four parameters are displayed on the PC screen every 15 s.

17.6.2 Aim

The aim of this project is to show how the X-NUCLEO-IKS01A2 expansion board can be used in a project.

17.6.3 Pin Configuration

X-NUCLEO-IKS01A2 expansion board pin configuration is shown in Fig. 17.6.

17.6.4 Program Listing

The program listing (program: **MEMS**) is shown in Fig. 17.7. The Mbed library **XNucleoIKS01A2** must be included in the program before the MEMS library functions can be used. Open a new Mbed session with program name **MEMS**, and click on **Import Library** and Click **Select Path** then select **MEMS** as the program name to include the library folder in the program space. The library is available at the following website:

https://os.mbed.com/components/X-NUCLEO-IKS01A2/

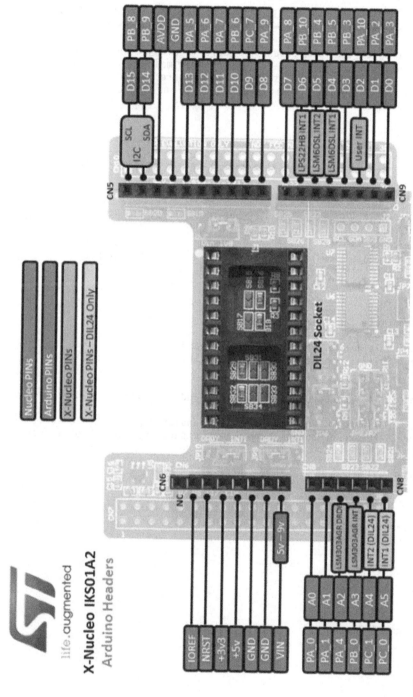

FIG. 17.6 X-NUCLEO-IKS01A2 expansion board pin configuration.

```
/*********************************************************************

       MEMS EXPANSION BOARD - HUMIDITY,TEMPERATURE,PRESSURE SENSOR
       ============================================================


In this project the MEMS expansion board is plugged on top of the
Nucleo-F411RE development board. The program measures the ambient
humidity, temperature, and atmospheric pressure. The Dew point is
calculated and all four parameters displayed on the PC screen
every 5 seconds

Author: Dogan Ibrahim
Date  : October 2018
File  : MEMS
*********************************************************************/
#include "mbed.h"
#include "XNucleoIKS01A2.h"

Serial pc(USBTX, USBRX);                            // Serial PC interface

#define SDA PB_9                                    // I2C1 SDA
#define SCL PB_8                                    // I2C1 SCL
#define LSM6DSLINT1 PB_5
#define LSM6DSLINT2 PB_4


//
// Instantiate the expansion board
//
static XNucleoIKS01A2 *mems_expansion_board =
XNucleoIKS01A2::instance(SDA, SCL, LSM6DSLINT1, LSM6DSLINT2);


//
// Humidity-temperature sensor (HTS211) and pressure (LPS22HB) handles
//
static HTS221Sensor *hum_temp = mems_expansion_board -> ht_sensor;
static LPS22HBSensor *press_temp = mems_expansion_board -> pt_sensor;


//
// Clear the screen
//
void clrscr()
{
    char clrscr[] = {0x1B, '[', '2' , 'J',0};
    pc.printf(clrscr);
}


//
// Home the cursor
//
```

FIG. 17.7   Program listing.

(Continued)

```
void homescr()
{
    char homescr[] = {0x1B, '[' , 'H' , 0};
    pc.printf(homescr);
}



int main()
{
    float temperature, humidity, pressure, DewPoint;
    float gamma, b, c;

    b = 17.67;
    c = 243.5;
//
// Enable the sensors used in the project
//
    hum_temp -> enable();                         // Enable humidity sensor
    press_temp -> enable();                       // Enable pressure sensor
//
// Get the humidity, temperature and pressure. Calculate the Dew Point
// and display all four on the PC screen every 15 seconds
//
    while(1)
    {
        clrscr();                                        // Clear screen
        homescr();                                       // Home cursor
        hum_temp -> get_temperature(&temperature);       // Get temperature
        hum_temp -> get_humidity(&humidity);             // Get humidity
        press_temp -> get_pressure(&pressure);           // Get pressure
        gamma = log(humidity / 100.0f);
        gamma = gamma + (b * temperature) / (c + temperature);
        DewPoint = (c * gamma) / (b - gamma);
        pc.printf("TEMPERATURE,HUMIDITY,PRESSURE,DEW POINT\n\r");
        pc.printf("=========================================\n\r");
        pc.printf("\n\r");
        pc.printf("Temperature = %5.2f C\n\r", temperature);
        pc.printf("   Humidity = %5.2f %%\n\r", humidity);
        pc.printf("   Pressure = %7.2f mba\n\r", pressure);
        pc.printf("  Dew Point = %5.2f C\n\r", DewPoint);
        wait(15);
    }
}
```

FIG. 17.7, CONT'D

At the beginning of the program header files **mbed.h** and **XNucleoIKS01A2.h** are included in the program. The PC serial interface and the interface between the expansion board and the development board are then defined. The handles for the humidity-temperature and pressure sensors are defined. Inside the main program the humidity-temperature sensor and pressure sensor modules are enabled. The program then enters an endless loop created using a **while** statement. Inside this loop the humidity, temperature, and pressure are read from the expansion board and stored in floating point variables

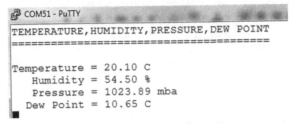

FIG. 17.8   Typical display.

humidity, temperature, and **pressure,** respectively. The dew point is calculated using the following formula by Magnus:

$$G = \text{Ln}(RH/100) + (b \cdot T)/(c + T)$$

$$\text{Dew Point} = c \cdot G/(b - g)$$

where Ln is the logarithm to base 2, RH is the relative humidity as a percentage, T is the ambient temperature in degree Centigrade, Dew Point is in degree Centigrade, and b and c are the following constants:

$$b = 17.67, \quad c = 243.5$$

For example, if $RH = 10\%$, $T = 25°C$ then the dew point is

$$G = \text{Ln}(10/100) + (17.67 \times 25)/(243.5 + 25)$$

or

$$G = -0.657$$

$$\text{Dew Point} = -243.5 \times 0.657/(17.67 + 0.657)$$

which gives, Dew Point $= -8.72°C$

A heading is displayed followed by all the four parameters. The display is repeated every 15 s. Fig. 17.8 shows a typical display from the program.

17.7 MULTIFUNCTIONAL EXPANSION BOARD (X-NUCLEO-IKA01A1)

This expansion board (Fig. 17.9) is based on operational amplifiers which can be used as an analog front end. The board contains a micropower and nanopower operational amplifiers. The basic features of this expansion board are as follows:

- seven predefined configurations: instrumentation amplifier, current sensing photodiode, ultraviolet current sensing, buffer, full wave rectifier, constant current LED driver, window comparator
- prototyping area on board
- Arduino UNO R3 connector compatible

FIG. 17.9   X-NUCLEO-IKA01A1 expansion board.

17.8 BLUETOOTH LOW-ENERGY EXPANSION BOARD (X-NUCLEO-IDB04A1)

This is a Bluetooth low-energy expansion board (Fig. 17.10) which interfaces with the Nucleo development boards via SPI bus. The board enables Bluetooth communication with other Bluetooth compatible devices. The basic features of this board are as follows:

- Bluetooth low-energy 4.0 master and slave compliant
- equipped with Arduino Uno R3 connector
- low-power consumption (7.3 mA RX and 8.2 mA TX)
- +8 dBm maximum transmission power
- −88 dBm receiver sensitivity

17.9 THREE-PHASE BRUSHLESS DC MOTOR BOARD (X-NUCLEO-IHM07M1)

This is a three-phase brushless DC motor expansion board (Fig. 17.11) based on the DMOS fully integrated driver L6230. The board enables three-phase brushless DC motor to be controlled from the Nucleo development boards. Overcurrent and thermal protection are provided and the driver is optimized for six-step and FOC algorithms. The basic features of this expansion board are as follows:

- 8–48 V operating voltage
- three-phase BLDC/PMSM motor driver
- 100 kHz operating frequency

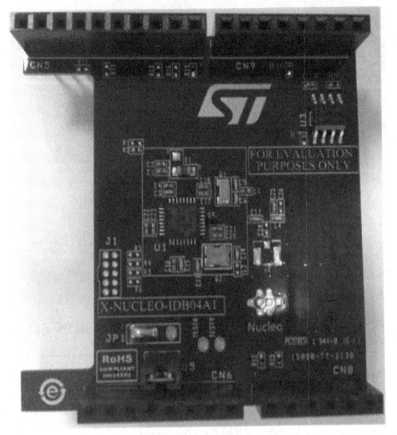

FIG. 17.10   X-NUCLEO-IDB04A1 expansion board.

- overcurrent and thermal protection
- motor current sensing
- Hall/encoder sensor connector and circuit
- potentiometer for speed regulation
- user LED

17.10 BIPOLAR STEPPER MOTOR DRIVER BOARD (X-NUCLEO-IHM05A1)

This is a bipolar stepper motor driver expansion board (Fig. 17.12) based on the L6208 driver. The board is compatible with the Arduino UNO R3 connector. Its basic features are as follows:

- 8–50 V operation
- up to 2.8 A rms phase current

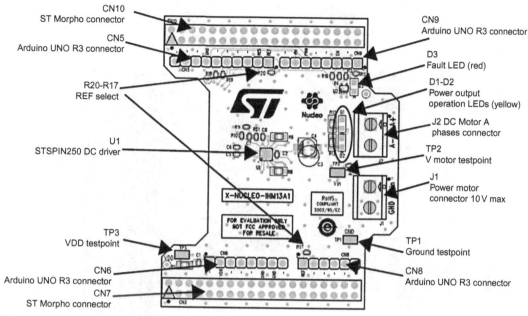

FIG. 17.11   X-NUCLEO-IHM07M1 expansion board.

- dual independent PWM current controllers
- full and half step drive
- fast and slow decay mode selection
- overcurrent and thermal protection

17.11 LOW-VOLTAGE STEPPER MOTOR DRIVER EXPANSION BOARD (X-NUCLEO-IHM06A1)

This is a low-voltage stepper motor driver expansion board (Fig. 17.13) based on the STSPIN220. The board can be operated with battery and is suitable for applications such as robotics, toys, and thermal printers. The board offers a maximum 1/256 microstep resolution and is Arduino UNO R3 connector compatible. The basic features of this board are as follows:

- 1.8–10 V operation
- up to 1.3 A rms phase current
- up to 256 step microstep adjustments
- overcurrent, short-circuit, and thermal shutdown protection
- current control with adjustable off-time

FIG. 17.12   X-NUCLEO-IHM05A1 expansion board.

17.12 BRUSHED DC MOTOR DRIVER EXPANSION BOARD (X-NUCLEO-IHM13A1)

This is a low-voltage brushed DC motor driver expansion board (Fig. 17.14) based on the STSPIN250 driver. The board is compatible with the Arduino UNO R3 connector. The basic features of this board are as follows:

- 1.8–10 V operation
- up to 2.6 A rms current
- overcurrent, short-circuit, and thermal shutdown protection
- current control with adjustable off-time

This board is described in more detail since two projects are given in this section using this board. The board component layout is shown in Fig. 17.15. Motor voltage (10 V maximum) is applied to connector J1, while the motor pins are connected to connector J2. **An external power supply should be used to provide power to the motor**. When power is applied through jumper J1 the red LED turns ON.

The following pins are used to control the board:

EN: Power enable input (logic HIGH to enable)
RESET: Reset
REF: Reference voltage for the current limiter circuit
PH: Logic input that determines the direction of rotation
PWM: Pulse Width Modulated input (logic HIGH to operate)

FIG. 17.13   X-NUCLEO-IHM06A1 expansion board.

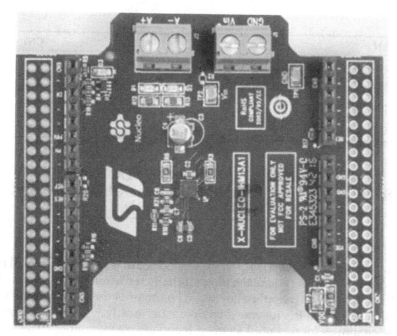

FIG. 17.14   X-NUCLEO-IHM13A1 expansion board.

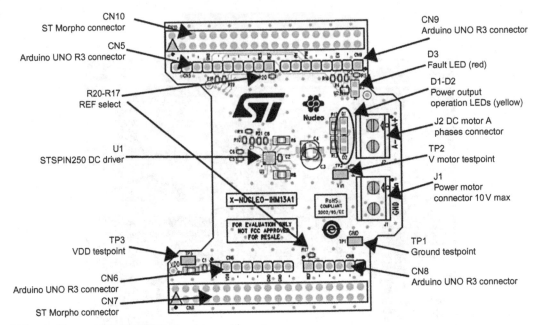

FIG. 17.15   X-NUCLEO-IHM13A1 component layout.

These pins are connected to the following Nucleo development board pins on the ST Morpho connector:

EN	PA_10	33	CN10
RST (STBY)	PC_7	19	CN10
REF	PA_0	28	CN7
PH	PB_10	25	CN10
PWM	PB_4	27	CN10

17.13 PROJECT 2—BRUSHED DC MOTOR SPEED CONTROL USING THE X-NUCLEO-IHM13A1 EXPANSION BOARD

17.13.1 Description

In this project an X-NUCLEO-IHM13A1 expansion board is used with a small brushed DC motor. This is a simple project where the motor initially starts rotating for 10s with 50% of its full speed. Then, full speed is applied for 10s. The motor is then stopped for 10s, and full speed is applied for 10s in the reverse direction. After this time the motor is stopped. The above process continues forever until stopped manually by the user.

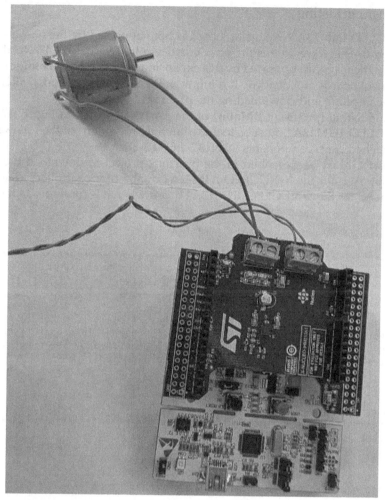

FIG. 17.16   Project components.

17.13.2 Aim

The aim of this project is to show how the X-NUCLEO-IHM13A1 expansion board can be used in a project to control the speed of a small brushed DC motor.

17.13.3 The Construction

The expansion board was plugged on top of the development board, a small DC motor was connected to screw terminal J2, and external +5 V power supply was connected to screw terminal J1. Fig. 17.16 shows the project components.

17.13.4 Program Listing

The X-NUCLEO-IHM13A1 expansion board is not supported by the Nucleo-F411RE development board. Therefore, in this project a board similar to Nucleo-L476RG development board is used which is fully supported by this expansion board. The pin configurations of the two development boards are identical. It is important that board Nucleo-L476RG must be selected before compiling and downloading the program.

The program listing (program: **BMotor**) of the project is shown in Fig. 17.17. The Mbed library **X-NUCLEO-IHM12A1** must be included in the program from the following website: https://os.mbed.com/components/X-NUCLEO-IHM12A1

Click **Import Library** as described in the previous project to download the library. Although this library is for the X-NUCLEO-IHM12A1 dual motor controller expansion board, it can also be used for the X-NUCLEO-IHM13A1 board when it is configured for single motor application.

```
/*******************************************************************

                    BRUSHED DC MOTOR CONTROL
                    ========================


This is a brushed DC motor control project. A small DC motor is
connected to the X-Nucleo-IHM13A1 expansion board. This board is
plugged on top of the Nucleo-L476RG development board. The program
rotates the motor in forward and backward diretions.

Author: Dogan Ibrahim
Date  : October 2018
File  : BMotor
*******************************************************************/
#include "mbed.h"
#include "STSpin240_250.h"

//
// Development board to motor controller board interface
//
#define PHA PB_10                    // PH pin
#define PHB PA_8                     // Not used
#define PWMA PB_4                    // PWM pin
#define PWMB PB_3                    // Not used
#define REF PA_0                     // REF pin
#define STBY PC_7                    // STBY pin
#define EN PA_10                     // EN pin


//
// The motor parameetrs must be initialized before starting the motor.
// Frequency of PWM input A maximum = 100,000Hz)
// PWM input B is not used
// Frequency of PWM for REF pin maximum = 100,000Hz
// Duty Cycle of PWM for REF pin is 0 to 100
// Dual bridge configuration, FALSE=mono, TRUE=dual
//
STSpin240_250_init_t init = {20000, 0, 20000, 50, FALSE};
STSpin240_250 *motor;
```

FIG. 17.17   Program listing.

(Continued)

```
int main()
{
    motor = new STSpin240_250(EN, STBY, PHA, PHB, PWMA, PWMB, REF);

    motor -> set_ref_pwm_freq(0, 10000);
    motor -> set_ref_pwm_dc(0, 50);
    motor -> set_bridge_input_pwm_freq(0, 10000);

    while(1)
    {
        motor -> set_speed(0, 50);              // 50% Duty Cycle
        motor -> run(0, BDCMotor::FWD);         // Run forward
        wait(10);                               // Wait for 10 sec
        motor -> set_speed(0, 100);             // 100% Duty Cycle
        wait(10);                               // Wait 10 sec
        motor -> hard_hiz(0);                   // Stop motor

        wait(10);                               // Wait for 10 sec
        motor -> set_speed(0, 100);             // 100% Duty Cycle
        motor -> run(0, BDCMotor::BWD);         // Run reverse
        wait(10);                               // Wait 10 sec
        motor -> hard_hiz(0);                   // Stop motor
        wait(10);
    }
}
```

FIG. 17.7, CONT'D

At the beginning of the program header files **mbed.h** and **STSpin240_250.h** are included in the program. Then, the interface between the expansion board and the development board are defined. Notice that although the library supports dual DC motors, in this project only a single motor is used and parameters for the second motor (PHB and PWMB are not used). Inside the main program the actual connections to the expansion board are defined with **motor** being declared a new class. Initially the PWM frequency is set to 10,000 Hz and the Duty Cycle to 50%. The remainder of the program runs in an endless loop created using a **while** statement. Inside this loop, the Duty Cycle is set to 50%, motor is configured to run in forward direction (BDCMotor::FWD) for 10 s. The Duty Cycle is then set to 100% and the motor runs a further 10 s. At this point the motor stops for 10 s. Then, the motor is configured to run for 10 s in backward direction (BDCMotor:: BWD) with 100% Duty Cycle. The above process is repeated forever until stopped by the user.

17.14 PROJECT 3—BRUSHED DC MOTOR SPEED CONTROL USING A POTENTIOMETER WITH THE X-NUCLEO-IHM13A1 EXPANSION BOARD

17.14.1 Description

In this project an X-NUCLEO-IHM13A1 expansion board is used with a small brushed DC motor. A potentiometer is connected to the Nucleo-L476RG development board. The speed of the motor is varied from 0% to 100% by moving the potentiometer arm.

17.14.2 Aim

The aim of this project is to show how the X-NUCLEO-IHM13A1 expansion board can be used in a project to control the speed of a small brushed DC motor.

17.14.3 Block Diagram

The block diagram of the project is shown in Fig. 17.18. The expansion board is plugged on top of the development board via the Arduino UNO R3 connectors.

17.14.4 Circuit Diagram

Fig. 17.19 shows the project circuit diagram. The arm of A10K linear potentiometer is connected to analog input PA_1 of the Nucleo-L476RG development board. The other two terminals of the potentiometer are connected to +3.3 V and GND pins of the development board.

17.14.5 Program Listing

The program listing (program: **PotMotor**) is shown in Fig. 17.20. Analog voltage of the potentiometer is read into variable **pot**. The value read is a floating point variable between 0 and 1.0,

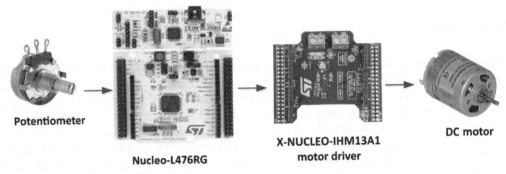

Potentiometer

Nucleo-L476RG

X-NUCLEO-IHM13A1 motor driver

DC motor

FIG. 17.18   Block diagram of the project.

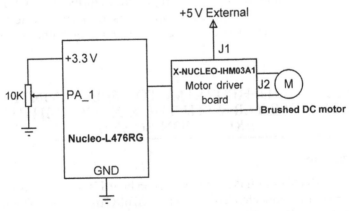

FIG. 17.19   Circuit diagram of the project.

where 0 corresponds to 0 V and 1.0 corresponds to the ADC reference voltage of +3.3 V. Variable **pot** is multiplied with 100 to convert it into Duty Cycle between 0% and 100%. This value is then used to control the speed of the motor such that as the potentiometer arm is rotated the motor speed changes from 0% to 100%. Make sure that you compile for the Nucleo-L476RG development board.

```
/***************************************************************
                BRUSHED DC MOTOR POTENTIOMETER CONTROL
                =========================================

In this project the X-Nucleo-Ihm13a1 motor driver expansion board
is plugged on top of the Nucleo-F476RG development board and a small
DC motor is connected to the expansion board. Also, a potentiometer
is connected to the development board. The speed of the motor is
controlled by the potentiometer

Author: Dogan Ibrahim
Date   : October 2018
File   : PotMotor
***************************************************************/
#include "mbed.h"
#include "STSpin240_250.h"
AnalogIn pot(PA_1);

//
// Development board to motor controller board interface
//
#define PHA PB_10               // PH pin
#define PHB PA_8                // Not used
#define PWMA PB_4               // PWM pin
#define PWMB PB_3               // Not used
#define REF PA_0                // REF pin
#define STBY PC_7               // STBY pin
#define EN PA_10                // EN pin

//
// The motor parameeters must be initialized before starting the motor.
// Frequency of PWM input A maximum = 100,000Hz)
// PWM input B is not used
// Frequency of PWM for REF pin maximum = 100,000Hz
// Duty Cycle of PWM for REF pin is 0 to 100
// Dual bridge configuration, FALSE=mono, TRUE=dual
//
STSpin240_250_init_t init = {20000, 0, 20000, 50, FALSE};
STSpin240_250 *motor;


int main()
```

FIG. 17.20   Program listing.

(Continued)

```
{
    float ain, DutyCycle;
    int DC;
    motor = new STSpin240_250(EN, STBY, PHA, PHB, PWMA, PWMB, REF);

    motor -> set_ref_pwm_freq(0, 10000);
    motor -> set_ref_pwm_dc(0, 50);
    motor -> set_bridge_input_pwm_freq(0, 10000);

    while(1)
    {
        ain = pot.read();                    // Read pot value
        DutyCycle = 100.0f * ain;            // Calc Duty Cycle
        DC = (int)DutyCycle;                 // As integer
        motor -> set_speed(0, DC);           // Set Duty Cycle
        motor -> run(0, BDCMotor::FWD);      // Run forward
    }
}
```

FIG. 17.20,   CONT'D

17.15 INDUSTRIAL DIGITAL OUTPUT EXPANSION BOARD (X-NUCLEO-OUT01A1)

This is an industrial output expansion board (Fig. 17.21) based on the galvanic octal power solid-state relay ISO8200BQ. The board provides eight relay-based outputs with LED indicators. The maximum output current per output channel is 700 mA.

The basic features of this expansion board are as follows:

- operating voltage 10.5–33 V
- galvanic isolated outputs
- 700 mA output current per channel
- thermal protection
- Arduino UNO R3 compatible connector

This board is described in more detail since a project is given in this section using this board. As shown in Fig. 17.21, an eight-way screw-type connector is provided at the bottom part of the board for making the output connections. Just above these connectors are the eight LEDs. At the top middle part of the LED a two-way screw connector (J2) is provided for applying external voltage (7–12 V) if the board is to be operated, for example, with a battery. An external 10.5–33 V DC voltage must be connected to the two-way screw terminal J1, located at the bottom right-hand corner of the board.

The X-NUCLEO-OUT01A1 expansion board has eight inputs, labeled IN1 to IN8 that correspond to the eight outputs. Jumper J4 selects the mode of operation as SYNC or

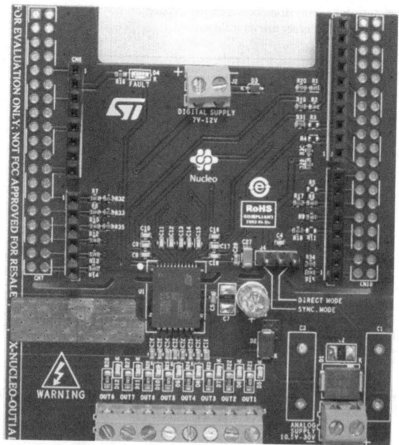

FIG. 17.21   X-NUCLEO-OUT01A1 expansion board.

DIRECT. For DIRECT mode this jumper must be connected to GND, and for SYNC mode the jumper must be connected to +3.3V. In DIRECT mode each input INi drives its corresponding output OUTi. The ISO8200BQ has an input buffer and a transmission buffer. The board is controlled by three signals named as LOAD, SYNC, and OUT_EN. OUT_EN must be set to logic HIGH to enable the channel outputs. LOAD input, when LOW, loads the input data into the input buffer. SYNC input, when LOW, copies contents of the input buffer to the transmission buffer and it can be used to synchronize the low-voltage side and the channel outputs on the isolated high-voltage side. In normal operation (DIRECT mode), LOAD and SYNC inputs are set to logic LOW, and the OUT_EN input is set to logic HIGH.

Fig. 17.22 shows the connections between the expansion board and the Nucleo development board. The pins of interest are as follows:

Expansion Board Signal	Nucleo Board Pin
IN1	PA_0
IN2	PA_1
IN3	PA_3
IN4	PA_4
IN5	PB_4
IN6	PB_5
IN7	PB_9
IN8	PB_8
LOAD	PC_0
SYNC	PC_1
OUT_EN	PB_0

A simplified block diagram of the X-NUCLEO-OUT01A1 expansion board is shown in Fig. 17.23.

17.16 PROJECT 4—CONTROLLING LEDs WITH THE X-NUCLEO-OUT01A1 EXPANSION BOARD

17.16.1 Description

In this project an X-NUCLEO-OUT01A1 expansion board is plugged on top of the Nucleo-F411RE development board. The LEDs on the expansion board are turned ON or OFF by entering commands in the form of hexadecimal numbers on the PC screen. For example, command 03 turns ON the two LEDs at the right-hand side of the LEDs, command 0F turns ON all four lower LEDs, command FF turns ON all the LEDs, and so on.

17.16.2 Aim

This is a very simple project. The aim of this project is to show how the X-NUCLEO-OUT01A1 expansion board can be used in a project with a Nucleo development board.

17.16.3 Circuit Diagram

The X-NUCLEO-OUT01A1 expansion board uses GPIO pin PA_3 for its channel 3 (IN3) input. But unfortunately PA_3 and PA_2 on the Nucleo development board are used as the USB serial port RX and TX signals of USART2, respectively (i.e., USBTX is same as PA_2 and USBRX is same as PA_3). If the boards are used as they are then channel 3 (OUT3) will not respond since its corresponding input pin IN3 cannot be controlled by the software. It is therefore necessary to redirect the USB serial port to another UART on the Nucleo board so that pin PA_3 becomes available as a general purpose digital pin.

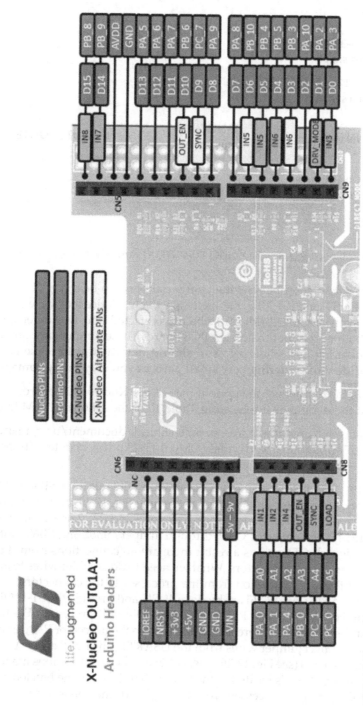

FIG. 17.22 Expansion board to Nucleo board connections.

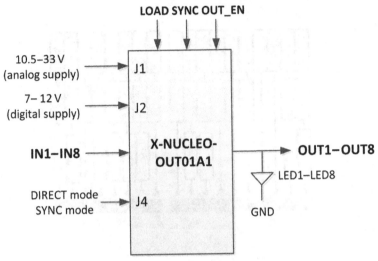

FIG. 17.23 Simplified block diagram of the X-NUCLEO-OUT01A1 expansion board.

The steps given below describe how the USB serial port can be changed to use USART 1 with pins PA_9 as the TX and PA_10 as the RX, respectively. Notice that these changes require two jumpers (also called solder bridges) to be disconnected and two other jumpers to be connected at the bottom part of the Nucleo PCB board. These jumpers are tiny and require very careful de-soldering and soldering. **You should not attempt to make these changes if you are not confident of soldering very small parts as you may easily damage the board**:

- Fig. 17.24 shows the jumpers at the bottom part of the Nucleo PCB board.
- The function of these jumpers are listed in Table 10 of the following STM User Manual:

  https://www.st.com/content/ccc/resource/technical/document/user_manual/98/2e/fa/4b/e0/82/43/b7/DM00105823.pdf/files/DM00105823.pdf/jcr:content/translations/en.DM00105823.pdf

- USART2 pins PA_3 and PA_2 of the processor are by default the USB serial RX and TX pins, respectively, for connecting to a PC. As shown in Fig. 17.25, these pins are connected to the Arduino Uno R3 pins D0 and D1 (or to the PA_3 and PA_2 on ST Morpho connector CN10), respectively, on connector CN9, through jumpers SB62 and SB63. But, these jumpers are by default OFF and as a result there are no connections from the processor PA_3 and PA_2 pins to the connectors. What we need to do is solder wires to jumpers SB62 and SB63 so that there is connection from the processor to the connectors.
- Jumpers SB13 and SB14 (Fig. 17.26) are by default ON and provide the ST-LINK USB serial link from the processor PA_3 and PA_2 pins to the ST-LINK processor. We have to make both these jumpers OFF so that there is no link between the processor PA_3 and PA_2 pins.
- We now have to connect jumper wires from the USART1 pins PA_9 and PA_10 to pins RX and TX of connector CN3 (see Fig. 17.26), respectively. These connections make USART1 to be the USB serial port. CN3 is located at the top left-hand side of the Nucleo board. Notice that PA_9 and PA_10 can be accessed from pin 1 of Arduino connector CN5 and pin 3 of Arduino connector CN9, respectively.
- The Nucleo board is now ready so that USART1 is the USB serial interface to a PC.

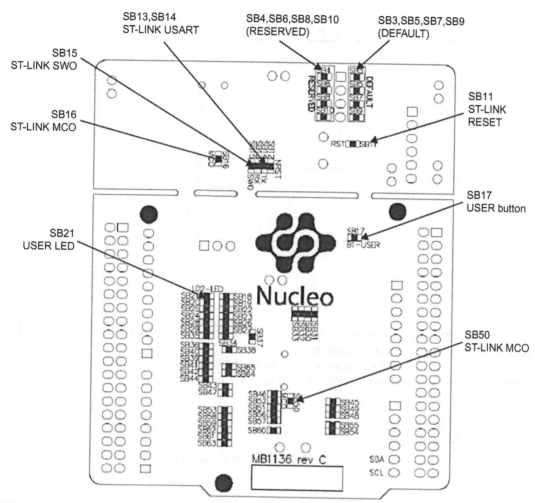

FIG. 17.24 Jumpers at the bottom part of the PCB.

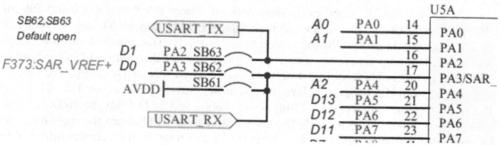

FIG. 17.25 Jumpers SB62 and SB63.

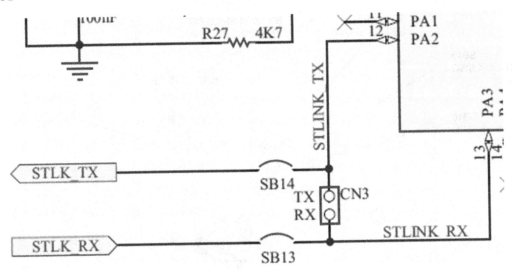

FIG. 17.26   Jumpers SB13 and SB14.

Remember to undo the above jumper changes if you wish to use USART2 (i.e., USBTX and USBRX) for USB serial interface.

An external 12 V DC voltage must also be connected to jumper J1 which is a two-way screw terminal located at the bottom right-hand side of the board.

17.16.4 The Construction

Fig. 17.27 shows the jumper wires for the USART1 connection.

17.16.5 Program Listing

The program listing (program: **OUT01A1**) is shown in Fig. 17.28. At the beginning of the program the **BusOut** statement is used to group the eight channels into a variable called **LEDS**. The USB serial PC interface is then defined where **PA_9** is the TX and **PA_10** is the RX pin, respectively. Inside the main program **LOAD** and **SYNC** are set LOW and **OUT_EN** is set HIGH so that the expansion board is ready. The remainder of the program runs in an endless loop where the user is prompted to enter the required LED pattern as a hexadecimal number. The required LEDs are then turned ON or OFF. Fig. 17.29 shows a typical output from the program. Here, for example, entering **ff** (binary: 1111 1111) turns all the LEDs ON, **02** (binary: 0000 0010) turns the second LED from the right-hand side ON, **06** (binary: 0000 0110) turns ON the second and third LEDs from the right-hand side, and so on. Notice that when an LED is turned ON the relay at the corresponding output channel is activated.

FIG. 17.27   Connections for USART1.

17.17 CR95HF NFC CARD READER EXPANSION BOARD (X-NUCLEO-NFC03A1)

This is an NFC card reader expansion board based on the CR95HF integrated circuit. The board can be used with a Nucleo board in security-based contactless tag applications, such as secure door entry, user identification, etc. The board provides 13.56 MHz interface to compatible NFC cards and communicates with the Nucleo boards through UART or SPI interface. The basic features of this expansion board are as follows (Fig. 17.30):

- Onboard 13.56 MHz inductive antenna supports ISO/IEC 14443 Type A and B, 15693, and 18092 communication protocols
- Detection, reading, and writing of NFC Type 1, 2, 3 and 4 tags
- Four general purpose LEDs
- Arduino UNO R3 compatible connectors

17.18 M24LR DYNAMIC TAG EXPANSION BOARD (X-NUCLEO-NFC02A1)

This is a dynamic NFC/RFID tag expansion board (Fig. 17.31) based on M24LR. It contains a 512 8 bit EEPROM in I^2C mode and 128×32 bit in RF mode. The board interfaces with the

```
/***********************************************************************

                X-NUCLEO-OUT01A1 EXPANSION BOARD LED CONTROL
                ================================================

In this project an X-Nucleo-OUT01A1 industrial expansion output board
is plugged on top of the Nucleo-F411RE development board. The LEDs
on the expansion board are controlled by sending commands from the PC
terminal. Foir example, sending FF (or ff) turnd ON all the LEDs,
sending 0f (or 0F) turns ON all the LEDs at the lower nibble positions
and so on.

Author: Dogan Ibrahim
Date   : October 2018
File   : OUT01A1
***********************************************************************/
#include "mbed.h"

BusOut LEDS(PA_0,PA_1,PA_3,PA_4,PB_4,PB_5,PB_9,PB_8);
DigitalOut LOAD(PC_0);                          // LOAD pin
DigitalOut SYNC(PC_1);                          // SYNC pin
DigitalOut OUT_EN(PB_0);                        // OUT_EN pin

Serial pc(PA_9, PA_10);                         // USB serial PC interface

//
// Main program. Enable the expansion board, read the required LED pattern
// and turn the LEDs ON and OFF as required
//
int main()
{
    int pattern;

    LOAD = 0;
    SYNC = 0;
    OUT_EN = 1;                                 // Enable outputs

    while(1)                                     // Do forever
    {
        pc.printf("\n\rEnter LED Pattern in Hex: ");
        pc.scanf("%x", &pattern);
        pc.printf("\n\rLEDs are turned ON/OFF...\n\r");
        LEDS = pattern;
    }
}
```

FIG. 17.28 Program listing.

Nucleo boards via the I²C interface and is Arduino UNO R3 connector compatible. The basic features of this expansion board are as follows:

- Up to 4 kbit memory
- Onboard 13.56 MHz inductive antenna
- Three general purpose LEDs
- Multiple boards can be cascaded for larger systems
- Self-powered or powered from the Nucleo boards

```
Enter LED Pattern in Hex: ff
LEDs are turned ON/OFF...

Enter LED Pattern in Hex: 02
LEDs are turned ON/OFF...

Enter LED Pattern in Hex: 06
LEDs are turned ON/OFF...

Enter LED Pattern in Hex: 07
LEDs are turned ON/OFF...
```

FIG. 17.29   Typical output.

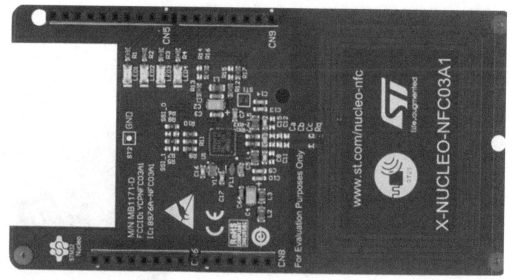

FIG. 17.30   X-NUCLEO-NFC03A1 expansion board.

17.19 RANGING AND GESTURE DETECTION SENSOR EXPANSION BOARD (X-NUCLEO-53L0A1)

This is a ranging and gesture detection sensor expansion board (Fig. 17.32) based on ST's Flight Sense (time-of-flight technology). This board provides an introduction to the ranging and gesture detection capabilities of the VL53L0X module. The board is compatible with the Arduino UNO R3 connectors. To allow the user to validate the VL53L0X in an environment as close as possible to its final application, the X-NUCLEO-53L0A1 expansion board is delivered with a cover glass holder in which three different spacers of 0.25, 0.5, and 1 mm height can be fitted below the cover glass in order to simulate various air gaps. Two VL53L0X satellites can

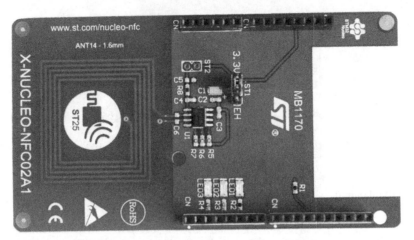

FIG. 17.31   X-NUCLEO-NFC02A1 expansion board.

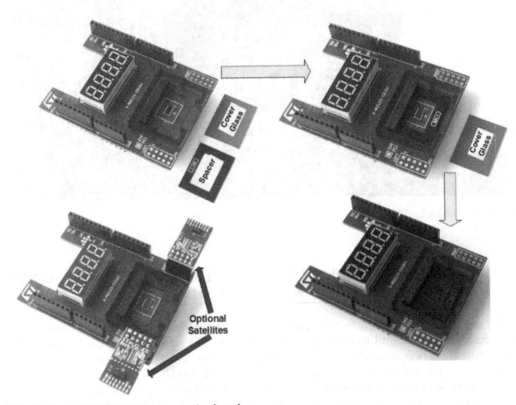

FIG. 17.32   X-NUCLEO-53L0A1 expansion board.

be connected using the two 10-pin connectors. The basic features of this expansion board are as follows:

- VL53L0X ranging and gesture detection sensor module.
- Accurate absolute ranging distance, independent of the reflectance of the target.
- Four-digit display, displaying the distance of a target from the ranging sensor.
- Basic gesture recognition application can be developed with a VL53L0X module.
- Cover glass.
- Two VL53L0X satellite boards.
- Two 10-pin connectors for VL53L0X satellite boards.
- Four-digit onboard seven-segment LED display.

This board is described in more detail since a project is given in this section using this board. The pin layout of this expansion board is shown in Fig. 17.33. The interface between the expansion board and the Nucleo development board is via I^{2}C.

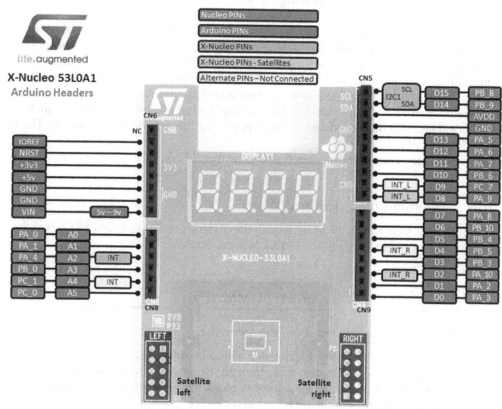

FIG. 17.33   Pin layout of the board.

17.20 PROJECT 5—MEASURING DISTANCE TO AN OBJECT USING THE X-NUCLEO-53L0A1 EXPANSION BOARD

17.20.1 Description

In this project an X-NUCLEO-OUT01A1 expansion board is plugged on top of the Nucleo-F411RE development board. The project measures the distance in millimeters to an object in-front of the expansion board and displays the result on the PC screen.

17.20.2 Aim

This is a very simple project. The aim of this project is to show how the X-NUCLEO-53L01A1 expansion board can be used in a project with a Nucleo development board.

17.20.3 Program Listing

Library **Xnucleo53L01A1** must be downloaded to the program space before the board can be used. This is available at the following website:

https://os.mbed.com/components/X-NUCLEO-53L0A1/

Create a new Mbed program, for example, called **Sonar**. Open the above website and click **Import Library**. Then select **Sonar** by clicking on **Target Path** to download the library to your program space. Fig. 17.34 shows the program listing. At the beginning of the program header files **mbed.h** and **XNucleo53L01A1.h** are included in the program and the I^2C interface between the expansion board and the development board are defined. The program initializes the expansion board and then enters an endless loop. Inside this loop the distance in-front of the expansion board is measured and is displayed in millimeters on the PC screen. The loop repeats after 1s delay.

A typical display from the program is shown in Fig. 17.35.

17.20.4 Suggestions for Additional Work

Modify the program in Fig. 17.34 to display the distance on the onboard LEDs (see the example program in the X-NUCLEO-53L01A1 Mbed website).

17.21 SUMMARY

In this chapter we have learned about the following:

- Information on commonly used Nucleo expansion boards
- Measuring humidity, temperature, and atmospheric pressure using an expansion board
- Controlling the speed of a small brushed DC motor using an expansion board
- Using the industrial I/O expansion board in a project
- Using the ranging and gesture detection sensor expansion board in a project

```
/**********************************************************************

                        SONAR PROJECT
                        =============

In this project an X-NUCLEO-53L01A1 sonar and gesture sensor expansion
board is pliugged on top of the Nucleo-F411RE development board The
program measures the distanxe in-front of the expansion board in mm and
then displays it on the PC screen every second.

Author: Dogan Ibrahim
Date   : October 2018
File   : Sonar
***********************************************************************/
#include "mbed.h"
#include "XNucleo53L0A1.h"

#define SDA PB_9                                // I2C interface
#define SCL PB_8                                // I2C interface
XNucleo53L0A1 *ExpansionBoard = NULL;

Serial pc(USBTX, USBRX);                        // Serial PC interface

//
// Clear the screen
//
void clrscr()
{
    char clrscr[] = {0x1B, '[', '2' , 'J',0};
    pc.printf(clrscr);
}


//
// Home the cursor
//
void homescr()
{
    char homescr[] = {0x1B, '[' , 'H' , 0};
    pc.printf(homescr);
}



//
// Goto specified line and column
//
void gotoscr(int line, int column)
{
    char scr[] = {0x1B, '[', 0x00, ';' ,0x00, 'H', 0};
    scr[2] = line;
    scr[4] = column;
    pc.printf(scr);
}
```

FIG. 17.34 Program listing.

(Continued)

```
int main()
{
    int stat;
    unsigned int DistanceToObject;
    DevI2C *i2c = new DevI2C(SDA, SCL);

    ExpansionBoard = XNucleo53L0A1::instance(i2c, PA_4, PA_9, PA_10);
//
// Initialize the board
//
    stat = ExpansionBoard -> init_board();
    clrscr();
    homescr();
    pc.printf("Dist=");
//
// Enter the loop to measure and display the distance
//
    while(1)                                          // Do forever
    {
        stat = ExpansionBoard -> sensor_centre -> get_distance(&DistanceToObject);
        if(stat == VL53L0X_ERROR_NONE)
        {
            gotoscr('0', '6');
            pc.printf("%ld mm    ", DistanceToObject);
            wait(1);
        }
    }
}
```

FIG. 17.34, CONT'D

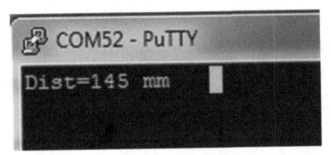

FIG. 17.35   Typical display from the program.

17.22 EXERCISES

1. Describe the advantages of using an expansion board in a project.
2. A motion MEMS expansion board is plugged on top of a Nucleo-F411RE development board. Write a program to read and display the acceleration on the PC screen as the board moves.

3. A motion MEMS expansion board and an LCD are connected to a Nucleo-F411RE development board. Write a program to read and display the relative humidity on the LCD every 5 s.

4. A brushed DC motor is connected to the expansion board X-NUCLEO-IHM13A1 which is plugged on top of a Nucleo-F411RE development board. Write a program to run the motor at 25% of its full speed.

5. An industrial digital output expansion board is plugged on top of a Nucleo-F411RE development board. Write a program to count up in binary on the LEDs every second.

Appendix

STM32 Nucleo-F411RE Pin Layout

The pin diagram is very important during the construction of all projects. Fig. A.1 shows the pin diagram of the STM32 Nucleo-F411RE development board.

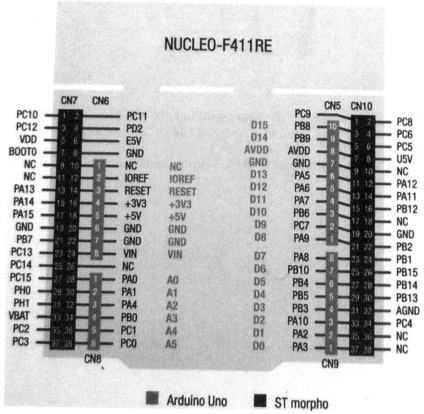

FIG. A.1   STM32 Nucleo-F411RE development board pin diagram.

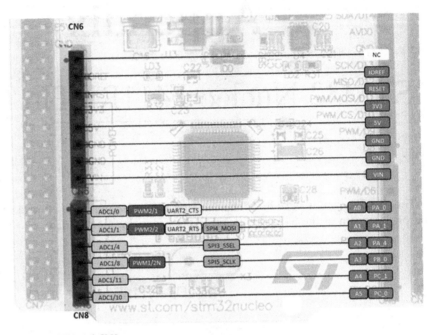

FIG. A.2 Arduino connector CN6 and CN8.

In addition to the pin diagram it is also useful to know the various functions of the pins. Figs. A.2 and A.3 show the Arduino connectors CN6, CN8 and CN5, CN9 pin functions, respectively.

Figs. A.4 and A.5 show CN7 and CN8 ST Morpho connector pin functions, respectively.

FIG. A.3   Arduino connector CN5 and CN9.

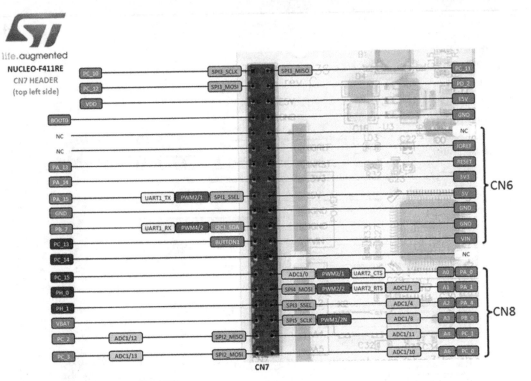

FIG. A.4 ST Morpho connector CN7.

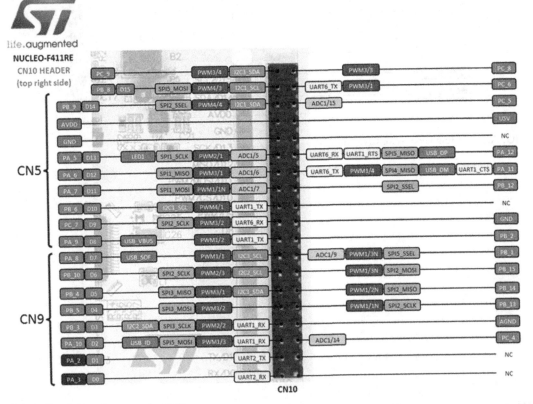

FIG. A.5 ST Morpho connector CN8.

Bibliography

Bear, C., Stallings, W., 2016. Wireless Communication Networks and Systems. Pearson.

Dam, B.V., 2010. ARM Microcontrollers: 35 Projects for Beginners. Elektor Press.

Horowitz, P., Hill, W., 2016. The Art of Electronics, third ed. Cambridge Univ. Press.

Ibrahim, D., 2000. Microcontroller Projects in C for the 8051. Newnes.

Ibrahim, D., 2006. PIC Basic Projects: 30 Projects Using PIC Basic and PIC Basic Pro. Newnes.

Ibrahim, D., 2010. SD Card Projects Using the PIC Microcontroller. Newnes.

Ibrahim, D., 2013. Practical Digital Signal Processing Using Microcontrollers. Elektor Press.

Ibrahim, D., 2014. PIC Microcontroller Projects in C: Basic to Advanced, second ed. Newnes.

Ibrahim, D., 2016. ARM Microcontroller Projects. Elektor Press.

Ibrahim, D., 2018. Programming with STM32 Nucleo Boards. Elektor Press.

Martin, T., 2013. The Designer's Guide to the Cortex-M Processor Family. Elsevier.

Norris, D., 2018. Programming With STM32: Getting Started With the Nucleo Board and C/C++. McGraw Hill.

Toulson, R., Wilmshurst, T., 2017. Fast and Effective Embedded Systems Design, second ed. Newnes.

Yiu, J., 2010. The Definitive Guide to the ARM Cortex-M3, second ed. Newnes.

Printed in the United States
By Bookmasters